Birds in Flight

written and illustrated by

John Kaufmann

WORLD'S WORK LTD
Kingswood Tadworth Surrey

For my mother

First published in Great Britain 1972 by
World's Work Ltd
The Windmill Press, Kingswood, Tadworth, Surrey
Printed by offset in Great Britain by
William Clowes & Sons, Limited
London, Beccles and Colchester
SBN 437 52120 6

Contents

1 Design for Flight

Every day, all around, birds are flying. Pigeons take off from pavements, flapping hard to pick up speed. Swallows skim low, turning quickly to snap up their insect prey. Falcons fold their wings in steep, swift dives to strike down other birds. Hummingbirds hover, standing still in mid-air on blurred wings to sip flower nectar. Hawks and gulls soar upward hundreds of feet in lazy spirals. With their long wings almost cutting the water, albatrosses streak low through the deep wave troughs of ocean gales.

Birds can be seen flying almost anywhere at any time. Yet we are so used to watching them flying with such ability and grace that we often take their flight for granted. Have you ever wondered how an eagle can float so high without flapping its wings, or how a blue tit manoeuvres so delicately when it grabs hold of a string of peanuts? Perhaps you too have felt a sense of amazement that living creatures can rise up against the heavy grasp of earth, relying completely on their own bodily skills to exploit the forces of the air, and spend hour after hour winging across the sky. If you look more closely you will find that birds in flight are both beautiful to watch and fascinating to study.

Imagine the situation of the bird in flight. There is no solid ground to hold it up. It does not have the buoyancy to float. A bird is an animal, made of bones, muscles, and feathers, heavier than the air through which it has to move. Thus it not only has to work to go forward, it has to work to hold itself up in the air against the downward pull of gravity acting upon its weight.

Accordingly, light weight is one important characteristic of any flying creature. The less weight a bird has to lift, the more easily it can fly. On the other hand, it must be built strongly enough to withstand the heavy strains of flight. An extremely light bird with a very frail skeleton would fold up like tissue paper in the first heavy gust of wind. Sturdy construction is also necessary to support the powerful flight muscles that pull its wings up and down. When these muscles contract, they pull hard upon different parts of the bird's skeleton, which must be strong enough to resist their force. Therefore, the physique of a bird is a precise balance among three things: great strength, plenty of power, and as little weight as possible.

Lightness of structure is especially marked among some of the larger birds. For instance, an American brown pelican, flying on five-foot wings and weighing twenty-five pounds, has a skeleton weighing only twenty-three ounces. Lightest of all, the subtropical frigate bird is a sea glider with a seven-foot wingspan and a body as small as a chicken's. Although it weighs between three and four

frigate birds

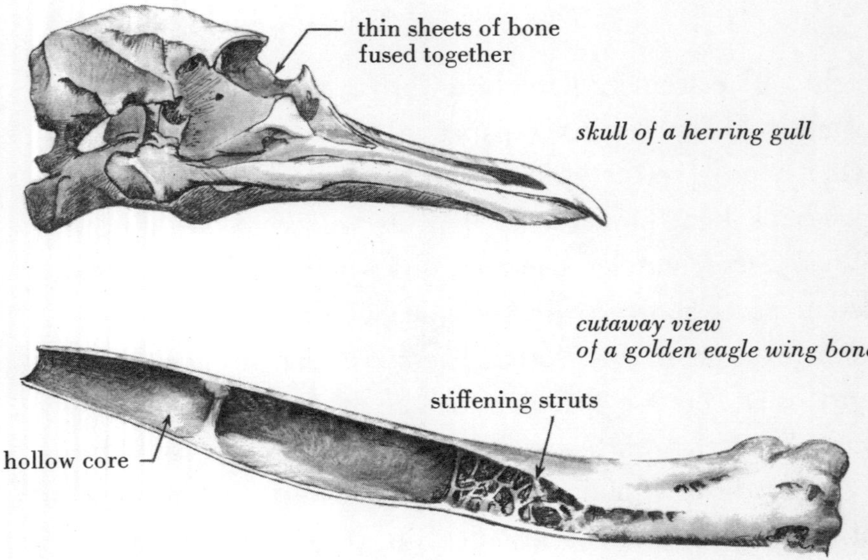

weight-saving features of bird skeletons

pounds, the frigate bird's skeleton accounts for a mere four ounces, which is less than the weight of its feathers.

The skeletons of birds maintain a balance between light weight and great strength in even their smallest parts. Bird bones are thinned down and hollowed out wherever possible. Their skulls are made of paper-thin sheets of bone, fused together and stiffened inside with thin rods. Bigger birds like eagles and vultures also have inner struts supporting their tubelike wing bones. Even the thin, tiny foot bones of many birds are hollow.

The skeleton of a herring gull is a good example. Instead of a simple, rounded rib cage, like that of man and other animals, the gull has a greatly extended breastbone, thin and deep like the keel of a yacht. Its ribs attach to the upper rear edge of the breastbone. They are long, thin, and flat, and they overlap each other in a light, flexible structure much like a basket woven from springy reeds. Two heavy shoulder bones rise up from the top of the breastbone and form sockets at their upper end. The inner ends of the wing bones rest and turn within these sockets.

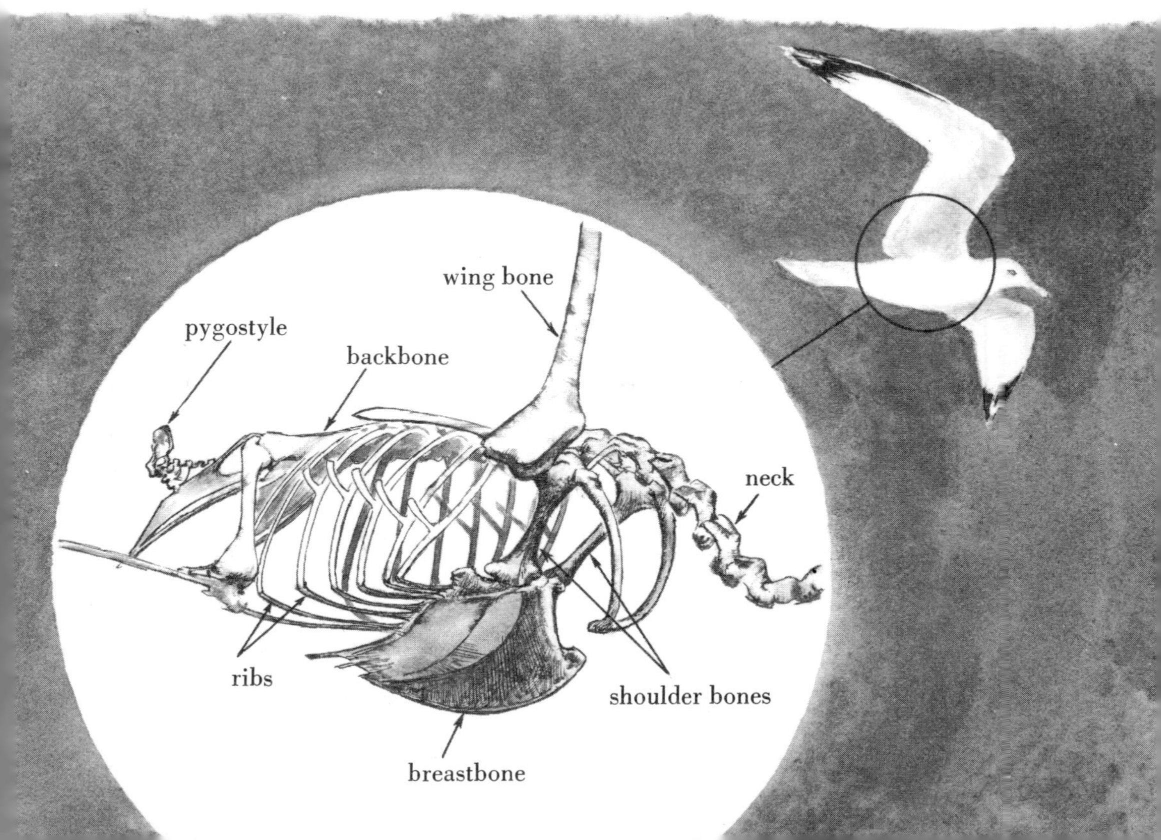

The breastbone, shoulder bones, and inner wing bones are the heaviest parts of the skeleton, since they carry both the inner pull of the bird's muscles and the outer pressures of the air acting on its wings. In addition, birds must withstand other forces that they encounter in their life. A falcon absorbs a great shock when it strikes a bird with its claws while flying at a speed faster than 100 miles an hour. A seabird resists extraordinary strains in a hurricane.

The wide, flat sides of the bird's breastbone provide lots of room for the attachment of its breast, or flight, muscles, whose outer ends connect to its wing bones. These muscles, the largest in a bird's body, are the power plant that enables the bird to flap its wings. The flight muscles of the strongest fliers and the hardest-working fliers are much larger than those of the weaker fliers such as herons. Because their kind of flying and their small size makes for inefficiency in flight, hummingbirds work very hard and need much more power for their size than any other birds. Thus the flight muscles of the smallest hummingbirds account for one third of their overall weight. Pigeons are very strong fliers, capable of flying in high winds when many other birds are grounded. Their flight muscles add up to about one half of their weight.

As a rule, the greater the depth of a bird's breastbone, and thus its fullness of chest, the more powerful its flight will be. The gull, which like vultures, eagles, and most hawks is primarily a glider rather than a flapper, has a shallow breastbone to carry its light flight muscles. Hummingbirds, swallows, swifts, falcons, ducks, geese, as well as sandpipers, plovers, and other shore birds, have deep chests and are all very strong fliers.

All these strong fliers have red flight muscles, an indication of a rich supply of red blood corpuscles capable of delivering plenty of body energy over extended periods of time. If a bird lacks a rich blood supply, despite having

hummingbird

a strong flapping flier
heavy flight muscles
deep breastbone

gull

a glider
light flight muscles
shallow breastbone

large flight muscles, its power will be limited to short bursts. Such is true of grouse, partridge, and quail, all deep-chested birds, having white flight muscles. When flushed by an intruder, they are capable of rocketing take-offs and very fast escape flights into nearby foliage. Under normal conditions, short rapid flights are suited to their way of life, and they have no need for sustained flying. However, if a dog flushes an American bobwhite quail several times in quick succession, the bird will exhaust its energy, and the hunter can walk over and pick it up.

At home, you probably have seen the white breast meat of chicken being carved for dinner. This meat is actually the flight muscles of a domesticated relative of an Indian jungle fowl, a strong but short-distance flier like a partridge. However, if you look at the leg meat of your bird, you will see the redness that indicates strong, constant physical activity. The colour is closer to that of the flight muscles of the American golden plover, which flies 3000 miles non-stop from Alaska to Hawaii during migration. The white muscles of the partridge provide momentary power, the red muscles of the plover provide power with endurance.

A bird's flight muscles are arranged in two sets, one to raise the wings, the other to lower them. The lowering muscles are heavier than the raising muscles, since the

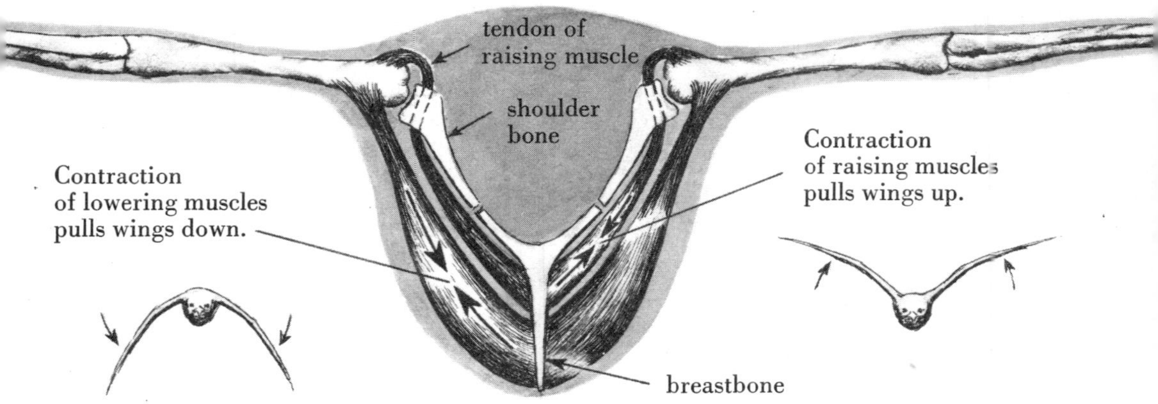

how the flight muscles operate

bird works harder when it flaps its wings down. For example, the lowering muscles of a thrush are about nine times heavier than the raising muscles. When the lowering muscles contract, they pull directly down on the wing. The raising muscles, however, have to be arranged to pull the wings upward when they contract. Therefore, they stretch upward from the breastbone, then connect to cord-like tendons that run through the shoulder bones and out to the wing bones. When the raising muscles contract, they pull down and the wings lift up. Although the flight-muscle arrangement may seem awkward, it works very smoothly. In fact, some hummingbirds are able to beat their wings as fast as eighty times a second.

A bird has a rigid, one-piece lower backbone. Other animals like cats, dogs, and horses have heavy, many-jointed spines. To hold their flexible backbones in correct position when walking or running, they need strong, heavy back muscles. The bird's backbone, however, is automatically fixed in the proper flying position, thus eliminating the need for weighty back muscles. To make up for the rigidity of their backbone, birds have a very flexible neck with fine muscular control, so they can use it for feeding, for nest building, and for preening their feathers.

The tail of most birds is also able to move freely. The light-weight tail feathers, which may make up much of its body length, are anchored to a fleshy knob on the pygostyle bone, which is located at the rear end of the backbone. The muscles controlling the tail are numerous and complex, enabling the tail to move in a variety of ways that help control the bird's manoeuvres.

In flight, the bones and muscles of a bird work together with the subtle smoothness and coordination that make any man-made machine seem crude by comparison. Powered by strong, elastic muscles pulling its airy frame, the bird sets forth on the ocean of the air, depending upon its flying performance for its survival in nature.

2 The Wing

High above a Florida beach, a turkey vulture circles slowly upwards, riding buoyantly on long wings spread motionless against the sky. From below, its wing tips seem to reach out like strong, dark fingers, to grasp the rising air. Low over a coastal pond, its sharp wings blurred, flapping fast and hard, a peregrine falcon charges a flock of sandpipers feeding on the shore. The sandpipers whirr upwards, one breaks from the flock, and a dark-feathered flash strikes. A faint thud, a puff of small feathers, and the peregrine streaks away, clutching its limp victim. The slow glide of the vulture and the speed dash of the falcon are two very different kinds of flight. However, both depend on the same basic device that gives birds their unequalled mastery of the air: the feathered wing.

All bird wings have certain structural parts in common. Think of the wing skeleton in terms of the parts of your own arms. Your shoulders support your upper arm bones, which connect at the elbows to your forearms. Wrist joints link your forearms to your hands, from which your fingers extend.

A bird's upper arm bones connect its wings to its body. Like your upper arm bones, although proportionately much smaller, they run between the shoulder and elbow joints. The bird's forearm sections, between the elbows and wrists, are called inner wings. They give most of the lift that supports the bird in flight. Gliding birds, such as the gull and the turkey vulture, have long inner wing sections.

The outer part of the wings, from the wrists to the wing tips, are called hand sections, or propellers, since they pull the bird forward when it flaps. Gliders like the gull have relatively small hand sections. Flappers, like the

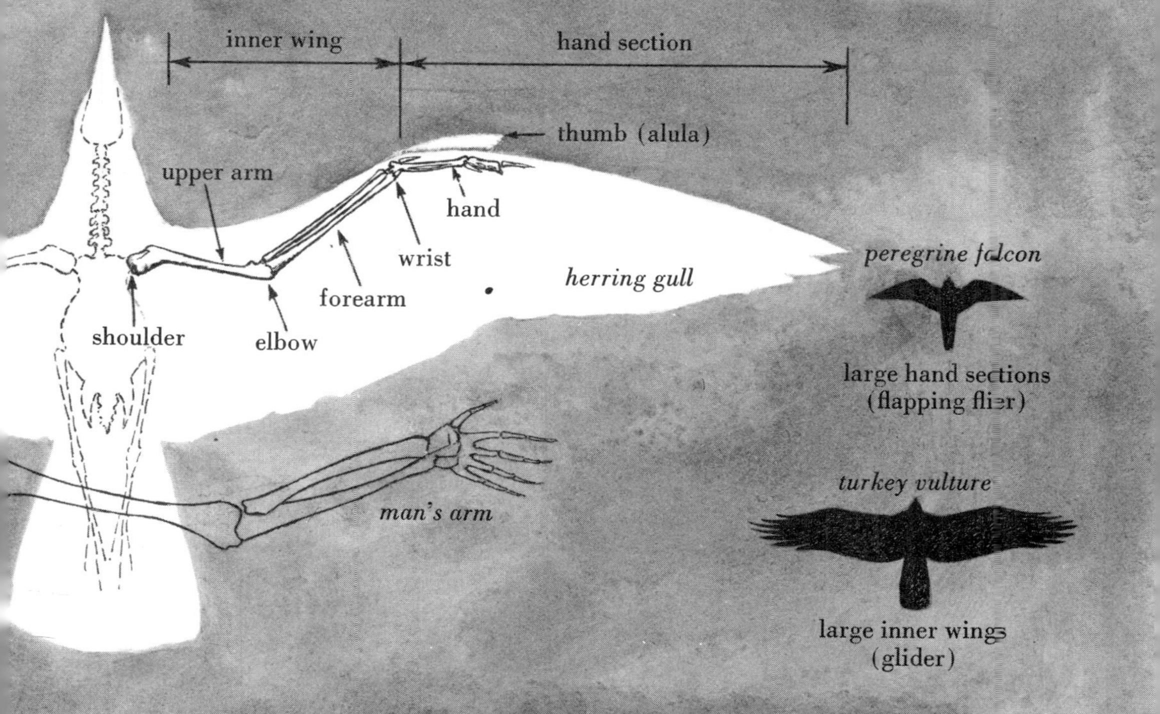

wing structure and proportions

peregrine, have large hand sections compared to their small inner wings. The thumb, attached to the front of the hand section, just beyond the wrist, is the little wing, or alula, which helps the birds to fly well at lower speeds, such as during takeoff and landing.

A bird's wings are extremely flexible. They can bend against the air striking them, and they can change their shape and position. In addition to moving its wings up and down, the bird also can control their degree of extension outward from its body. Extensor muscles attached at the inner ends to the body skeleton connect to long, cordlike tendons, which run out through each wing, cross the elbow and wrist joints, and end at the hand bone. Tightening the extensor muscles pulls the tendons, straightens the joints, and extends the wings. The wing tendons of a glider like the gull are highly developed, allowing it to hold its wings spread in the gliding position for long periods with little effort.

The extensor control permits birds to use a variety of wing positions, from sharply bent, or flexed, to fully spread. Other muscles can pull the wings forward or back, rotate them at the shoulder joints, and raise the front (leading) edge and rear (trailing) edge, either together or separately. The bird can fan its wing tips out wide or retract them. It can control the shape and position of each wing and alula section individually.

Overall the wing is designed in the shape of a "Z." This ingenious design enables the bird to fold its wide spreading wings into a sleek, curving shape that fits tight against its body. Who could possibly guess that the body of the wandering albatross, so short, plump, and compact on the ground, could conceal a magnificent wing, which

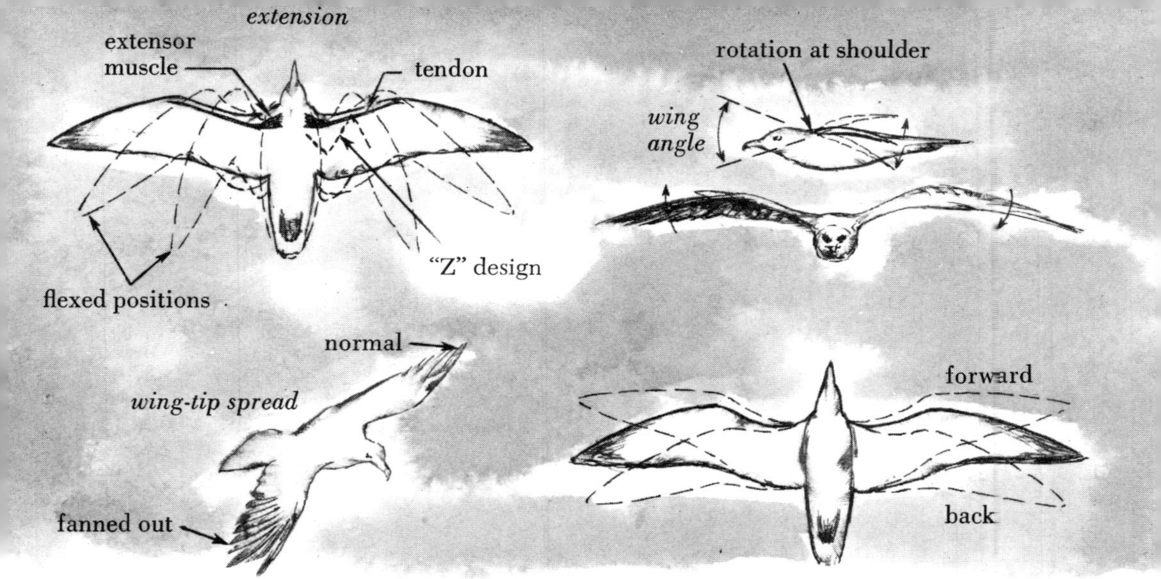

wing controls

opens up to a span of ten feet and more in the air? The wing folding of a gull, a thrush, or a sparrow is not as sensational, but is no less amazing. The flexibility, strength, and muscular control of the bird wing places it among the most precise and highly refined structures in all nature. While able to withstand the full force of a gale, it also can respond to the slightest variations in the currents of the air.

Bird wings come in many sizes and shapes. A bird's wing design depends mainly upon the habitat in which it lives and on its feeding habits. Birds that live in and fly through woods and undergrowth tend to have short, broad wings. Ground-feeding birds, like the woodcock, quail, and partridge, can scoot for cover quickly through dense foliage with little danger of injuring their very short, rounded wings. Even the wing-tip feathers are bent back to prevent snagging. Their wing design shows they are creatures of the quick escape. They fly fast for short distances and alight as soon as they reach safety.

Short wings also provide thrushes and blackbirds with excellent manoeuvrability as they dodge among vines and shrubs. Sparrows search among grasses for seeds on the ground; most warblers scour tree branches for their insect food. Accordingly, both sparrows and warblers have short wings that enable them to move easily through tight spots and give them the quick agility needed to reach insects within hidden niches in foliage. The sparrowhawk has rounded wings that permits it to dart and twist through

trees and bushes, and to turn sharply in pursuit of its prey. At night, woodland owls thread their way through dark trees on blunt, silent wings.

Out in open country a long, narrow pointed wing may be more advantageous. Basically it is a more efficient shape in flight. If its habitat and feeding habits permit, a bird can move faster and fly farther with less effort on long wings with sharp tips. A falcon does not have to make tight turns in confined places, so it has no need for the short wings of the sparrowhawk. To attack, the peregrine falcon makes one long, swift pass. If it misses, its speed and momentum are so great that it usually will not be able to come about for another pass before the prey escapes. Instead the peregrine moves on, to strike again elsewhere.

Long-winged swallows and swifts do not pick among grasses and trees for insects. They too are birds of the open spaces, which sweep the sky for their flying insect food. Thus their wings are well shaped for their environment. Despite their high-speed wing design, swallows can twist abruptly and almost halt in midair to intercept and snap up elusive insects. They and a number of other fast fliers have surprisingly good manoeuvrability. Birds like swallows, swifts, and terns spend much of their normal day in fast, continual flapping flight, so they depend heavily upon the efficiency of the pointed wing for their endurance.

Shorebirds like sandpipers and plovers live and feed on wide open mud flats and beaches, where long, pointed wings are highly suitable. Since many shorebirds make long, nonstop migrations over the sea, they must have the great endurance made possible by their efficiently shaped wings. The fact that an American golden plover can beat its wings for two days and two nights, flying at about fifty miles per hour across twenty-four hundred miles nonstop from Labrador to South America, while burning only about two ounces of its body fat for fuel, never ceases to amaze the aircraft designers. Most birds that migrate very long distances have long, pointed wings, including the graceful and powerful arctic tern, whose yearly round trip of up to 22,000 miles between the arctic and antarctic makes it the greatest bird traveller of all.

The conditions under which birds fly also have a strong influence on their wing design. Birds that do much gliding in the wide spaces of sea and shore usually have long, narrow, and sharply tipped wings. Gannets, shearwaters, and albatrosses must contend with the steady, strong winds that blow across open stretches of ocean. As a result, they have wings capable of gliding at high speed. Since they live and feed where there are no obstructions and no tight spots to fly through, these gliders have little need for the greater manoeuvrability and slower speeds made possible by the shorter, broader wing.

Over land, gliding birds must move more slowly. Land gliders use rising currents of air caused by certain heated

areas of the ground below. These heated currents are limited to certain locations, they vary with changing weather conditions, and they often may be weak and widely scattered. A fast glider, covering distance rapidly, could easily pass such currents by, and even if it detected them they would be ill-suited to its style of flight. The broad wing design permits very large lifting surfaces with a minimum wing span. It allows most hawks, as well as eagles, buzzards, and vultures, to glide slowly and search out local rising currents carefully. A slow land glider like the buzzard can detect even the slightest upward movement of air. Its broad wings enable it to turn quickly in a tight circle, to find and hold the best flying position within the current, and to work its way easily from one current to another when necessary.

Thus, each bird's wings are designed to function under the conditions of the air in the terrain where it lives. They must help the bird in feeding, nesting, migrating, and all its other activities. From the closely crowded thicket, where the robin flutters from one spot to another, to the open expanse of the sky at sea, where the long-winged albatross skims the surface in wide, sweeping glides, each bird uses its own specially shaped wings to live and thrive within its own particular environment.

3 The Wing in Motion

A buzzard passes by. What holds it up? What happens when its wings, or the wings of any bird, move through the air? Look at the form of the buzzard's wing. It is not just a flat surface covered with feathers, but a curved and streamlined shape. This special shape, called an aerofoil, is thicker at the leading edge and thins down towards the trailing edge. It is streamlined for the same reason a boat is pointed at the bow and curved inwards towards the stern. Streamlining enables a boat to travel through water and a bird to slip through the air as smoothly and easily as possible. As a result, less power is required to move either the boat or the bird.

The aerofoil is curved to make the air through which it moves act in certain ways. A normal, undisturbed airflow is like a smooth stream, running in parallel lines. When the buzzard's wing moves through it, the airflow is pushed apart, upwards and downwards, across the top and bottom surfaces. Since the top surface is curved more than the bottom, the upper airstream is deflected farther from its course and has a longer way to travel to reach the trailing edge. In order to travel across the wing in the same amount of time, the upper airstream must move faster than the lower airstream.

Whenever a fluid medium such as air flows faster past an object, the air pressure on the object decreases. Thus, in the case of the wing, the faster-moving air on top presses down with less force than the slower air below presses up. This situation causes a suction on the top surface, pulling the wing upwards, while the positive pressure of the air below pushes it in the same direction. The two pressures together make up a vertical, upward force called lift, and lift is what enables birds to fly. The upward suction on

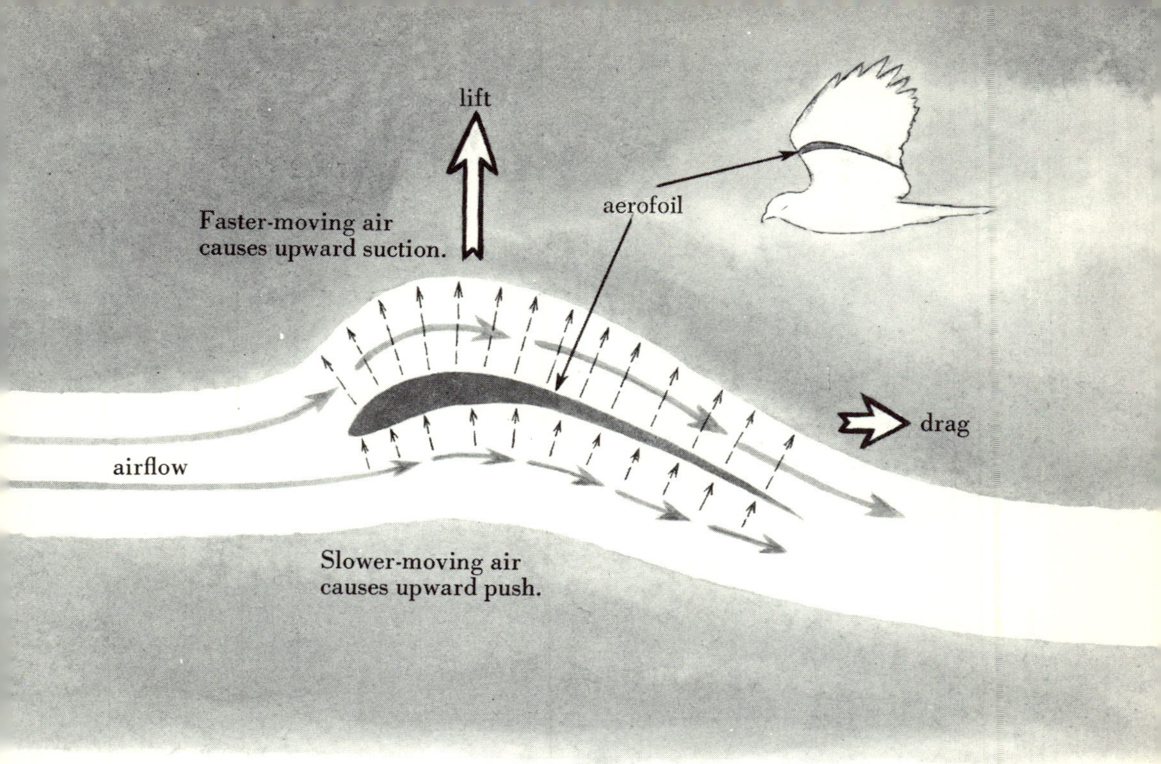

how the wing creates lift

top, rather than the force pushing from below, gives the buzzard's wing most of its lift.

To increase its lift, the buzzard can tilt its wings up in front so they meet the airflow at a higher angle. The angle between an aerofoil and the airflow is called the angle of attack. As this angle becomes greater, the curved upper aerofoil surface deflects the air passing over it farther upwards. Thus, the distance that the upper air must travel increases, the air speeds up and causes greater suction on top, and the amount of lift is increased. Since the air below now strikes the bottom surface at a more direct angle, it flows more slowly and also adds to the lift.

The buzzard can only tilt its wings to a certain point, however, and still gain more lift. While its lift continues to increase, the buzzard's wing encounters more resistance. You have felt air resistance holding you back when you have ridden fast on a bike. The backward force caused by air resistance is called drag. Although the aerofoil's streamlining keeps the drag as low as possible, this resistance still increases rapidly as the angle of attack becomes greater. In fact, if a buzzard raises its wing angle two degrees, it doubles the amount of its lift, but causes four times as much drag.

Drag builds up quickly, because at higher angles of attack the air above the top surface is deflected so far that it can no longer pass smoothly over the wing. Its flow begins to break away from the surface and forms eddies that become more and more turbulent as the angle increases. Finally the drag increases so much that it cancels out any small gain in lift. Usually at about fifteen degrees angle of attack the turbulence becomes so great that the airflow completely separates from the upper surface of the wing. Drag goes way up and lift disappears. The wing can no longer hold its high angle, so it drops down abruptly. This complete loss of lift is called stalling.

A stall is a very serious matter to a buzzard or any other bird, since it lands and takes off at a high angle of attack. If its wings should stall and drop suddenly during takeoff or landing, when it is close to the ground, a bird could suffer a serious injury.

The bastard wing, or alula, permits the buzzard to fly at high angles of attack without stalling. By extending and

how the alula works

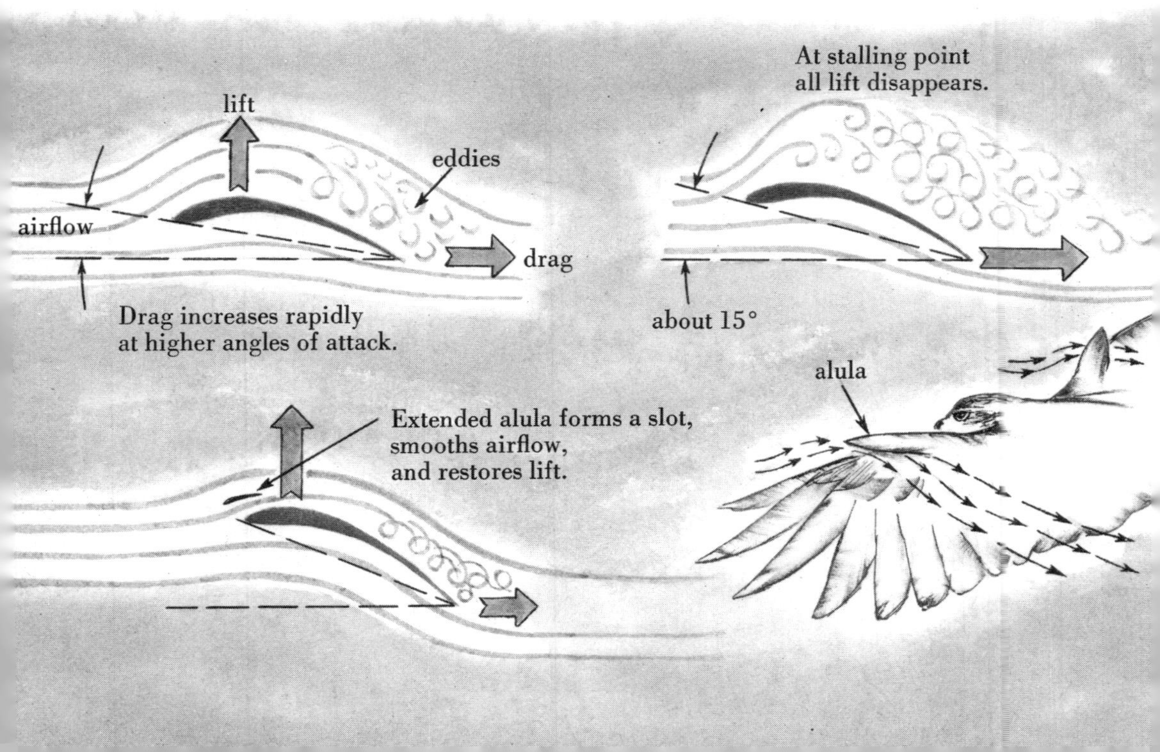

raising the alula, it forces a stream of air to flow through the narrow opening and across the upper wing surface. Like a river that flows fastest where it is narrowest, the air speeds up as it passes through the alula slot. This added flow not only smooths out the turbulence on top of the wing and restores lift, its extra speed reduces the air pressure above and increases lift where it is most needed.

The alula is especially important to those birds that fly at low speeds in their daily activities. Among these species are woodland birds such as nightingales, warblers, thrushes, and blackbirds, which have to manoeuvre through underbrush and trees. The alula's high lift also helps partridge, grouse and pheasants to make steep takeoffs. Slow land gliders depend upon their alula sections, too. For example, the turkey vulture can tilt its wings to the extreme angle of twenty-eight degrees before reaching the stalling point. Vultures do not carry their prey, but eagles, hawks, and owls do. They get the extra lift they need to carry their victims back to the nest from the high-wing angles made possible by the antistall action of the alula.

Bird wing design further takes into account another kind of drag, which occurs when air passing below the wing curls up over the trailing edge, drawn there by the low pressure on top. This upward whirl of turbulent air

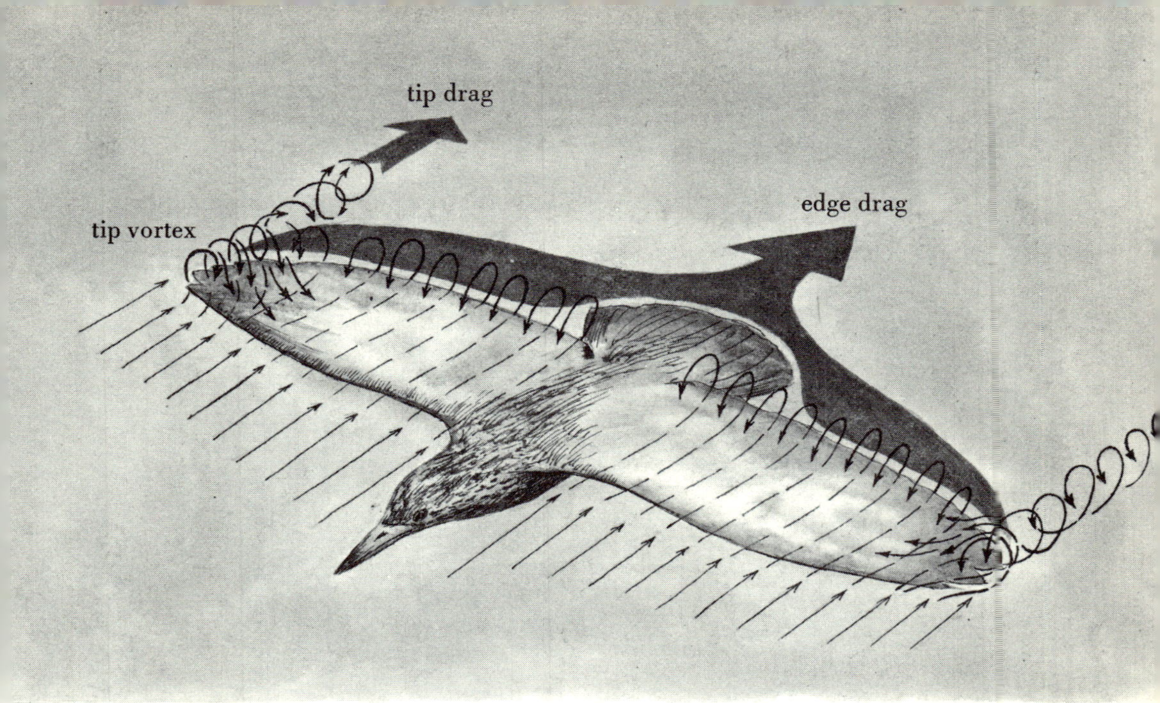

behind the wing is especially strong at the wing tip, where it swirls up around the edge in what is known as a tip vortex. A tip vortex not only causes drag and destroys lift at the tip alone, it spills over the upper wing surface and interferes with the airflow there too. To keep this interference as low as possible, many bird wings are markedly narrowed and lengthened so that the tips on either side of the body are widely separated. In this way the largest possible area between the tips remain unaffected by the turbulence. Thus birds with the longest, narrowest wings are often the most efficient fliers.

Among large birds, the wandering albatross has the longest, narrowest wings. Its wings span more than ten feet, while their width from the front edge to the rear edge is only about nine inches. Such slender proportions spread the wing tips apart so their drag affects a very small portion of the whole wing. In addition to the long, narrow wing shape, birds like albatrosses, frigate birds, and terns have sharply pointed wing tips. A pointed tip

advantages of long, pointed wings

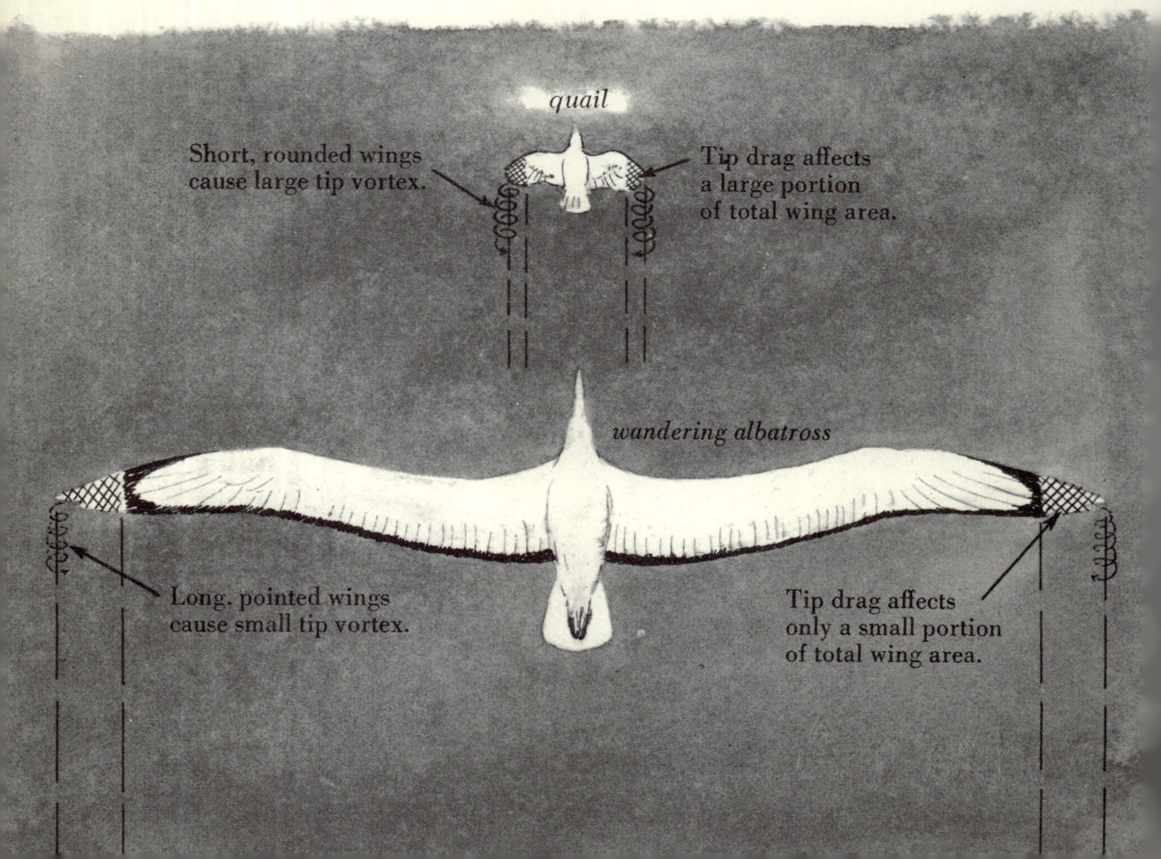

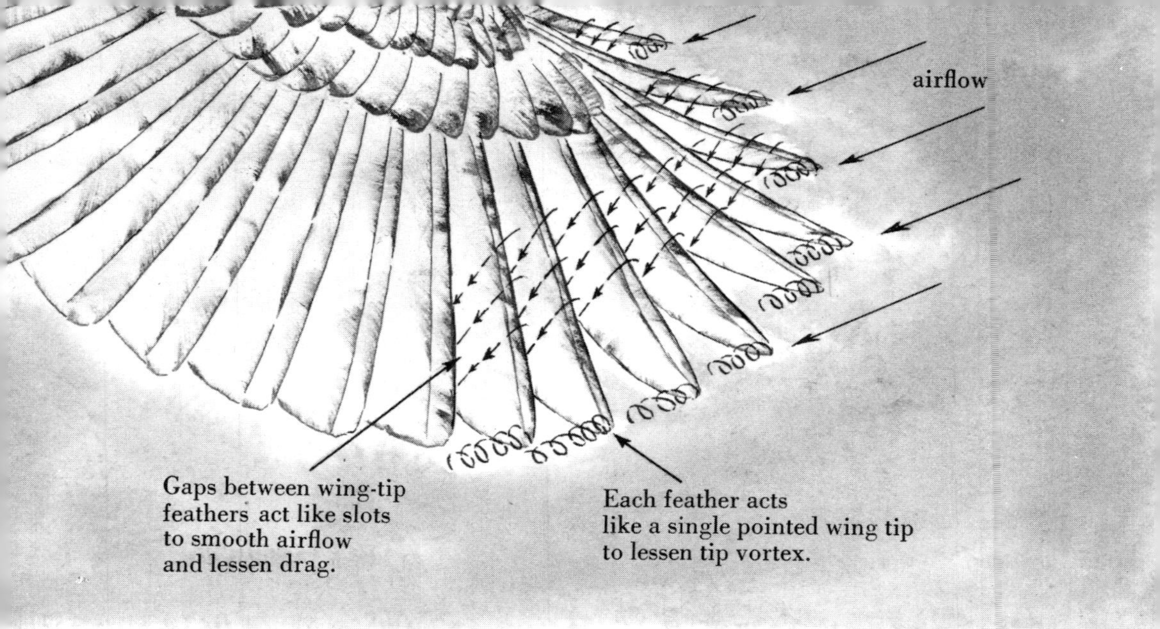

how notching makes a more efficient wing

reduces the tip vortex and exposes as little hand-section surface area as possible to the vortex turbulence, causing the least drag and making a highly efficient wing.

Not all birds without long, narrow, pointed wings are poor fliers, however. Many species of birds whose life habits require the shorter, broader wing have been provided with a notched wing tip, which is an alternative way to achieve efficiency. When a broad wing tip is notched into a series of smaller tips, each one acts like a narrow, pointed wing and gives much lift with little drag. Notching

Andean condor

prevents the buildup of a large tip vortex along the broad edge and the excessive drag that would result from it. In addition, the notches act like rows of slots to smooth the airflow over the wing tip at slow speeds and high angles of attack. For this reason the condor, the crane, and other large, slow fliers have deeply notched tips.

Trailing edge and tip turbulence affect the size as well as the shape of bird wings. Compared to a big bird like a Canada goose, a sparrow has more of its wing area disturbed by edge and tip drag. Therefore, the sparrow is equipped with proportionally larger wings than the goose to make up for its drag losses. The goose is more efficient in flight simply because it is bigger and so has relatively smaller wings.

Aerofoil shapes are designed to operate best within certain limits of speed and for particular kinds of flight. Slower fliers like crows, hawks, and vultures have thicker, more deeply curved aerofoils than faster fliers like falcons, swifts, sandpipers, and ducks. Birds with aerofoils curved deeply for high lift do not have to fly fast, since their wing curvature provides enough lift at low speed.

On the other hand, a bird like the swift has to move fast to make up with speed what it lacks in aerofoil curvature. Its wings, like those of all fast fliers, operate most efficiently at low angles of attack. The thin, flat aerofoil at high speed and a low angle of attack gives as much lift as a thick, curved aerofoil at low speed and a high angle.

In fact, an increase in airspeed causes much more than a proportional increase in lift. For example, if a swift doubles its angle of attack from two to four degrees, it doubles its lift. However, by doubling its speed, say from thirty to sixty miles per hour, the swift gets four times as much lift. Of course, most birds cannot double their flight speed easily, but even smaller increases in speed add to lift very quickly. Thus many ducks, all fast fliers, have small wings. Their high speed makes up for their thin, flat aerofoil shapes and for the comparatively small area of their lifting surfaces.

Again the buzzard passes by. Its wings slip through the thin medium of the air, tilted upwards to meet its flow, to separate and shape its currents. To the eye, not a feather ruffles. There is no sign of anything going on. But silently the bird uses invisible forces to create lift, overcome its weight of bone, muscle, flesh, and feather, and to sustain its wings in flight.

4 Feathers

As light as a feather. When you watch a small, puffy feather floating and spinning lightly on a breeze, you can appreciate that expression. The lightness, strength, and flexibility of feathers make them the most important factor in birds' unique mastery of flight. Insects and bats fly, but on very different wings. Only birds have feathers, and largely because of them they can fly faster, higher, and farther than any bat or insect. Feathers give the bird its

down feather

semiplume feather

finished shape and enable this heavier-than-air animal to convert its muscle power and weight into the forces of flight.

Feathers provide a bird with its clothing, keeping it warm in winter and cool in summer. They serve as patterned markings so birds of the same species can recognize each other at mating time, while the protective colouring of feathers acts as camouflage to help conceal the birds from predators. Feathers themselves do not transmit any feeling of cold, heat, or pain. They give the bird a strong, protective outer coat, which is repaired every six months or year, when old feathers are discarded and a new set grows. If a bird loses a feather accidentally, a new one automatically grows.

Like reptile scales and human fingernails, feathers are made of keratin, a tough, lightweight material. They grow from anchoring points within the bird's flesh, supported only by their own strong, flexible shafts. There are several main types of feathers, each one designed to do a certain job. Down feathers have short shafts and are fluffy, like ruffled hair. Their soft, spreading plumes and many tiny enclosed air spaces make them fine insulators. They form an inner lining under the bird's outer feathers and keep its body temperature constant despite winter cold and summer heat.

Semiplume feathers have longer shafts supporting spreading plumes that spread out on both sides. They serve as firmer padding for the outer feathers and also provide insulation. Both down and semiplume feathers act as a shock-absorbing lining, which helps protect the bird's light skeleton from injury if it should strike something in flight.

Vaned feathers are the most important of all, since they form the outer surfaces of the bird. Their name comes from their thin, flat vanes, which are formed by two areas called webs, one on each side of the shaft. Look closely at

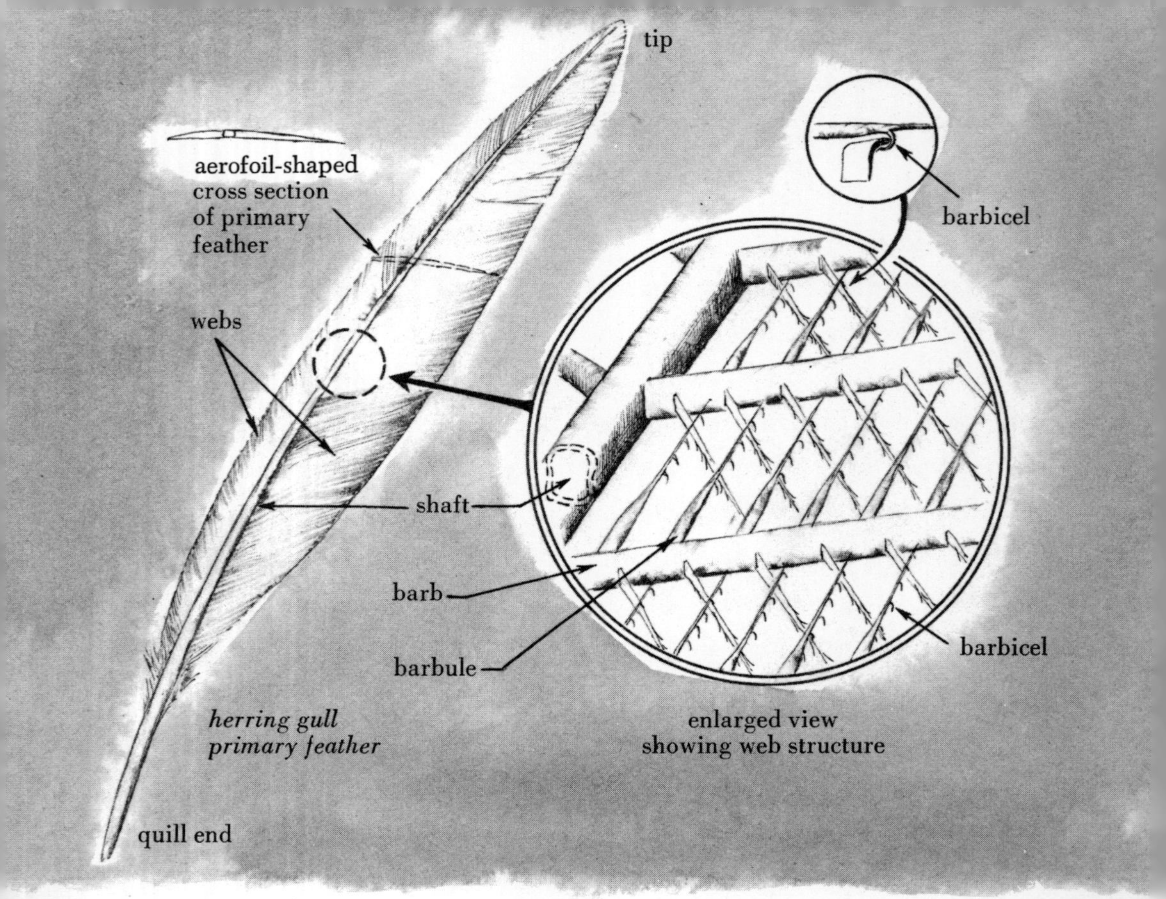

structure of vaned feather

a typical vaned feather from the hand section of a gull's wing. The shaft is long, hollow, and tapered outwards from the base, or quill end, to the pointed tip. The vane itself is shaped in a very precise curve, which narrows down towards the tip.

The webs are not solid, flat sheets, but are made up of many separate, very lightweight parts that latch together. Thin rods, called barbs, attach to the shaft sides. Each web is made up of hundreds of barbs, side by side. The barbs are held together flat within the web by overlapping fasteners called barbules, which are so small that a single feather may have more than a million. The barbules in turn overlap and attach with tiny hooks called barbicels. If the barbs become separated accidentally and the web splits apart, the bird simply zips the barbs together again by preening them with its bill until the barbicels hook. When you see a bird preening, it is getting its feathers back into good flying shape and also rubbing them with oil secreted by the preen gland at the rear of its back. This oil waterproofs the feathers and is especially important to waterbirds such as ducks, as well as seabirds like petrels and shearwaters. Waterlogged feathers would add excess weight, lose their shape, and stick together in flight.

The most important vaned feathers are the flight feathers. The term is usually used to refer only to the larger feathers of the wing and does not include those of the tail. However, the large tail feathers are also important in flight, since they too form the flying surfaces that help to support and control the bird. The wing flight feathers, as they might better be called, are anchored within the patagium, a tough, elastic membrane that runs outwards from the body to the hand section. The feathers are grouped according to which part of the wing skeleton they are joined to.

First, the primary feathers, whose outer ends form the wing tips, are supported by the hand bones. When the wings are folded, the hand sections of some birds, such as swifts, terns, and frigate birds, extend beyond the tail and overlap. In such cases, the primary feathers are very long. Next, the secondary feathers extend backwards from the forearm sections, or inner wings.

wing feather groups

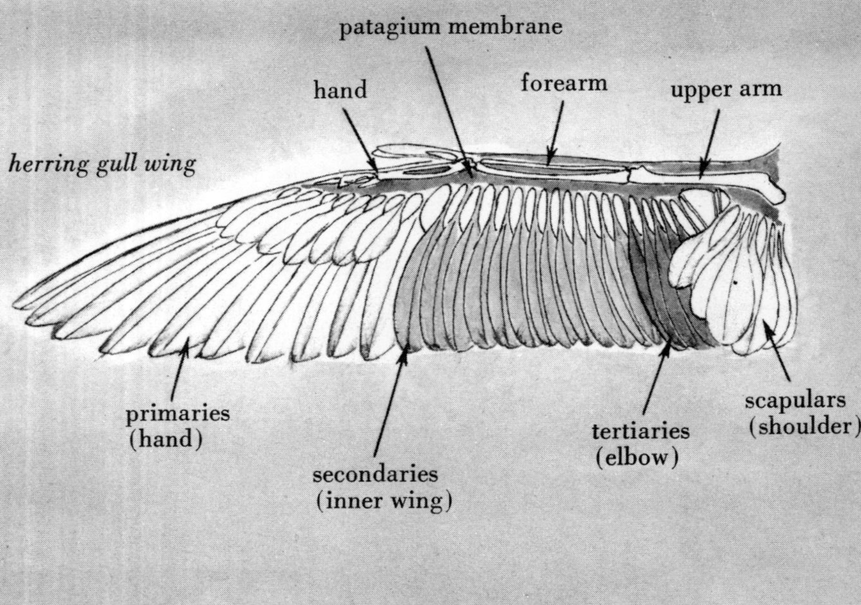

Last are two small groups, the tertiary and scapular feathers. The tertiary feathers are fastened to the inner, or elbow, ends of the forearms, and the scapular feathers attach to the upper arm section. Both groups fill in the gap between the bird's body and its main flight feathers, and they are very flexible so they can tuck away easily when the wings fold.

In all these groups the feathers overlap each other for added support against the pressure of the airflow, which tends to separate and distort them. The overlap effect is like that of a fanned-out hand of playing cards. Each card by itself bends easily, but when overlapped they are stiff.

If you find a discarded flight feather, bend it back as far as it will go. Some feathers curve back very far without the shaft collapsing. The strength and flexibility of the shaft and vane are especially important in the primary feathers. In flight, a bird's wings frequently and rapidly change position, and the airflow strikes them from many different angles. Since the wing tips are at the outer end of any wing movement, the primaries undergo the greatest changes in position, bending the most and carrying the heaviest strains. Their strength and flexibility allow them to twist and bend constantly, in order to maintain the proper angle to the airflow, without snapping off.

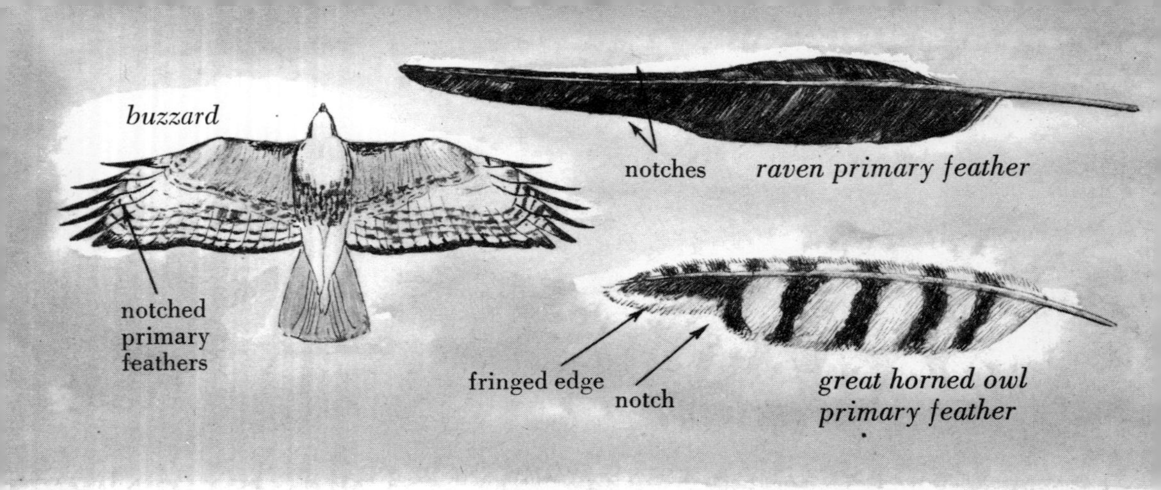

The primaries of many notched wings are also notched themselves. For example, the front and rear webs of a buzzard's primaries are narrowed down, or notched, for a portion of their length. This structure enables each feather to get the greatest possible support by overlapping the others, yet at the same time the narrow tip provides the greatest efficiency in flight. Many other birds, for example the raven and the great horned owl, have notched primary feathers too. Owl primary feathers also are fringed at the edges. This feature softens the airflow across the vanes, muffling the wing sound so the owl can move silently through the dark to surprise its prey.

The primary feathers of certain other birds serve very special purposes. The American nighthawk, a dusky relative of the swift and our nightjar, has long, pointed wings, which propel it with nimble, bounding strokes after flying insects. During the male's courtship flights, the nighthawk performs before his mate-to-be with steep, spectacular dives. At the bottom of each dive he pulls up abruptly and snaps his wings open wide. The airflow shoots through his primaries, making them vibrate in a loud *swoosh*.

Two sharply thinned-down primaries in front of each hand section on the male broad-tailed hummingbird make a high-pitched rattling buzz as he dives rapidly and pulls up before the female. The female's wings are not notched, so she flies with a normal humming sound. Three sharply cut-down primaries on the American woodcock's hand section cause a high whistling noise in flight, and play a part in the male's courtship display.

primary feathers used in courtship flights

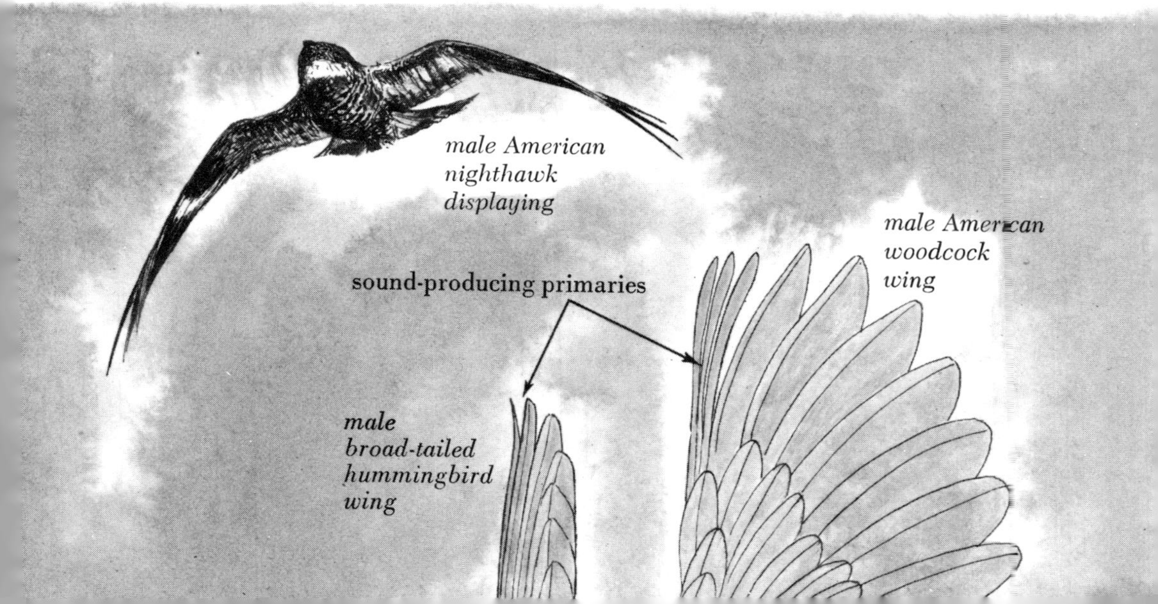

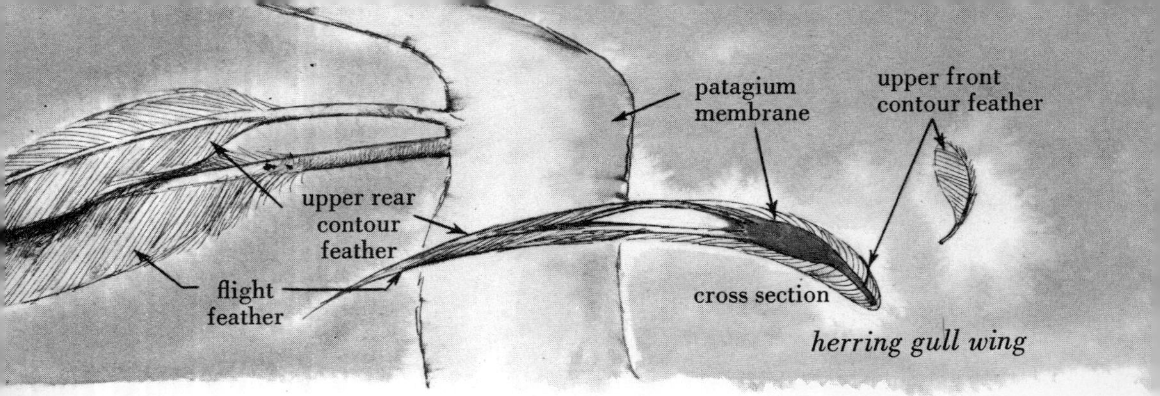

The contour feather is another kind of vaned feather that shapes the wing. Much of the wing's streamlined aerofoil is formed by contour feathers, which are smaller, thinner, and less structurally strong than the flight feathers. On a gull's wing, for example, each secondary feather has a contour feather directly above it, giving the aerofoil its upper rear curvature. The upper front surface is formed by contour feathers attached to the patagium membrane at right angles to its surface, then curved sharply backwards, where they overlap each other like springy roof tiles for extra support. Other contour feathers complete the aerofoil shape. Thus the combination of flight and contour feathers streamline the wing while keeping it flexible and light in weight.

Additional contour feathers sleekly cover the bird's head, neck, breast, and other body parts, allowing it to move smoothly through the air with the least drag. In fact, much of the outer form of a bird is actually a light shell of contour feathers, curved and placed in exactly the right way.

Sometimes flight feathers become worn down and ragged. This condition occurs from much hard flying, especially during migration. Birds go through a periodic process called moulting, when they shed their old feathers and grow new ones. The contour feathers of many species of birds moult twice a year, changing between the bright colours and strong patterns of springtime and the drab plumage of winter. The flight feathers of most species, however, moult only once a year. Thus even the flight feathers of long-distance migrants like the arctic tern must last a whole year, through thousands of miles of flying. Understandably their edges often become frayed before they are replaced.

Moulting can have a marked effect on flight. Many water-birds, such as ducks, geese, swans, and divers, shed all their primary feathers at once and remain flightless for a few weeks until new ones grow. They survive this period,

order of moult of the flight feathers

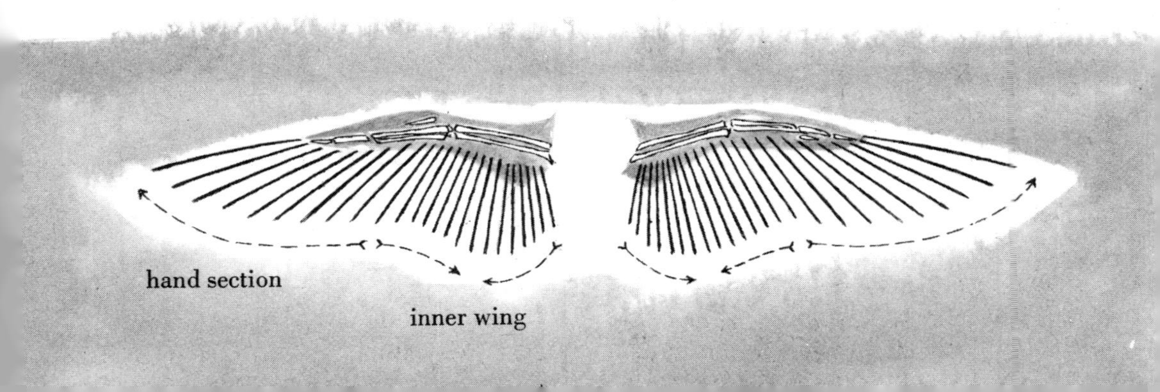

because they live and feed on or near the water, where they can escape danger by diving or by hiding in the reeds. Land birds cannot moult this way, since flying is essential to their feeding and their ability to escape predators. Flightless, they would be easy game for hawks, falcons, snakes, and other enemies. Instead, they moult just a few flight feathers at a time, so that their ability to fly is only slightly affected.

Flight feathers do not moult at random, but in a regular order. In many birds the primary feathers moult from the wrist outwards to the wing tip, while the secondaries moult from both ends of the inner wing toward the centre. To keep the wings balanced from side to side, with equal lift and drag, two matching primaries fall from each wing at the same time, leaving gaps for the new ones. The next matching set of primaries will not fall out until the previous set has almost finished growing. Thus, despite a complete feather overhaul, the bird can keep on flying.

Feathers shape the bird. They are the link between the animal beneath their quills and the air beyond. With great strength they bend and yield to blasts of pressure from the air. With flexible sensitivity they respond to the slightest fluctuations of its currents.

5 Gliding and Soaring

It is midmorning, with a light breeze blowing across open fields. A turkey vulture appears over the edge of a nearby pine wood, searching for food. It glides along, rocking slightly now and then on widespread wings, its long, dark primaries bent up against the air.

Gliding is the simplest form of flight. Just by holding its wings in a spread position, the vulture moves through the air. Its glide can be compared to the movement of a

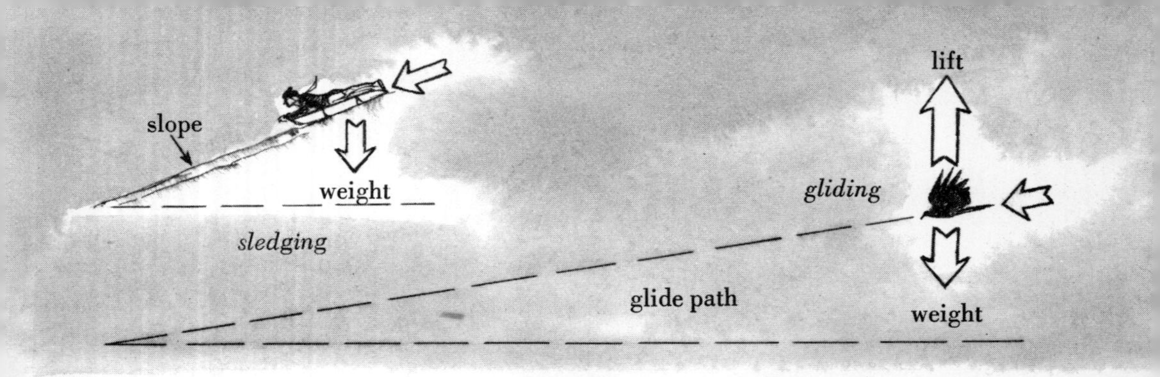

sledging and gliding

sledge going downhill. Gravity pulls the sledge's weight down the slope, overcomes the friction of the runners, and moves the sledge forward. Gravity also pulls the vulture's weight down, moves its wings through the air, and causes them to lift. The combined result of the two forces, downward gravity and upward, forward lift, moves the vulture ahead.

The more lift the vulture has, the better it will glide. Like other birds, such as seagulls, albatrosses, eagles, and most hawks, that depend mostly upon gliding, the turkey vulture has large inner wings and relatively small hand sections. The inner wings of any bird create much less turbulence than the hand sections, so they give more lift and less drag. In addition, the hands are either pointed or notched, as well as thinned down toward the tips. They do not have the full aerofoil shape like the inner wing, so their lift is less for that reason as well. Therefore, when a vulture spreads its wings to glide, most of its lift comes from the inner wing areas.

The wandering albatross has the longest wings of any glider, and the inner sections are greatly elongated. The tips are so far apart and so small that their drag is kept to an absolute minimum, and the inner wings make up most of the entire wingspan. However, a long span is not the only indication of large inner wings. Condors, for example, also have huge inner wings, because of their broadness from leading to trailing edge, although the wings are shorter than those of the albatross. Thus, the total wing area rather than the span is what determines the size of the inner wing.

Two main things determine how well a bird can glide. The first is the efficiency of its wings in flight, the proportion or ratio between their lift and drag. For example, an albatross has a lift-to-drag ratio as high as thirty to one. For every thirty ounces of vertical lifting force there is

different inner wing designs of two large gliders

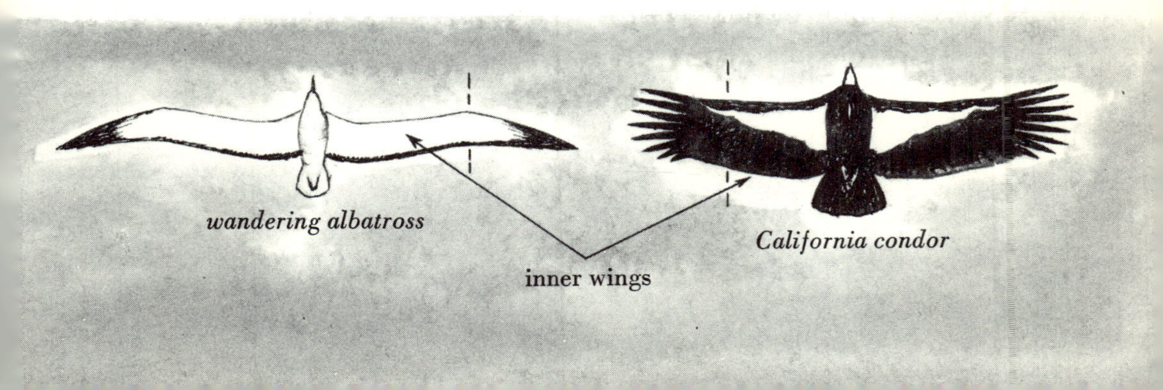

wing loading

only one ounce of resistance to its forward movement. A high lift-to-drag ratio means that a glider wastes little power in overcoming drag, and that most of its power can go into moving forwards and lifting itself.

The other important factor is how much weight the bird's wings must carry. The amount of wing area a glider has, compared to its weight, is called its wing loading. This figure can be given in square inches per ounce. The turkey vulture has about 674 square inches of wing area to carry its 85 ounces of weight, so its wing loading is 7.9 square inches per ounce. The golden eagle has a heavier wing loading, with about 1020 square inches of wing area for 164 ounces of weight, or 6.8 square inches per ounce. A heavier wing loading is indicated by a smaller figure, since a smaller wing area is used to lift the same weight, causing a heavier load. A glider with a lighter wing loading needs to use less of its lift to overcome its weight, so it can move forwards in a longer, flatter glide.

The path of a gliding bird is a sloping line, since it travels so many feet forward for each foot of height lost. The gliding ability of birds can be easily determined by comparing the steepness of their glide slope. Seagulls, which are especially fine gliders, travel more than twenty feet forward for each foot they drop. Albatrosses do almost as well while eagles have a glide slope of about sixteen to one.

Another way of measuring a bird's gliding ability is by noting its sinking speed, the amount of height it loses in a given time. Sinking speed is determined primarily by wing loading, so that lighter, larger gliders descend more slowly

glide slope

than do smaller, heavier ones. For instance, the turkey vulture has a lighter wing loading than the smaller black vulture, so its sinking speed is only two feet per second compared to the black vulture's rate of two and six-tenths feet per second.

In order to fly for some period of time without flapping, a glider must find extra power to sustain it in the air. Otherwise, its glide slope would bring it sooner or later down to the ground. When vultures, hawks, seagulls, and other gliders hold the same altitude or climb higher without flapping, they are not merely gliding, but are using the power of rising air currents. If a current of air is moving upwards with enough force, a bird can glide downwards within that current, but still rise higher above the ground. This use of rising air currents is a special kind of gliding called soaring.

One kind of soaring uses updrafts, currents deflected upwards when they strike mountain ridges, sand dunes, lines of trees, bridges, or tall buildings. When the turkey vulture gliding over the fields wants to gain height, it heads for a line of trees. Approaching on the downwind side, it slips up over the trees, positions itself within the updraft flowing over the treetops, and suddenly rises in the deflection current. When gulls soar straight and level

along the length of a bridge, they are riding on the currents deflected upwards from the windward side. When gulls hang motionless in the air behind a moving boat, they are soaring on the currents pushed upwards by the boat.

Probably the most spectacular use of deflection currents is found among the hawks migrating in autumn along mountain ridges, such as in the Kittatinny Mountains of Pennsylvania, U.S.A. When conditions are best, with a northwest wind crossing the ridge, hundreds of hawks a day can be seen skimming by on set wings, getting a free ride towards the South. Each autumn thousands of people come to watch the hawk migration at the famous Hawk Mountain Sanctuary in this area.

Another more important kind of soaring is practised widely by vultures, hawks, gulls, and other gliders over land on sunny days. When the turkey vulture has gained height on the deflection current of the tree line and wants to rise still higher, it looks for a ploughed farm field or a road. Gliding along, flapping as little as possible, it searches for an ascending current of air.

Suddenly, as if lifted by some invisible hand, it rises almost vertically, wheels around, and starts gliding in a tight, upward spiral. In a few minutes it is hundreds of feet above the ground, still rising and scanning the land below for food. By now some gulls have spotted the turkey vulture, so they hurry over to join it. Before long, the flock of gulls is circling far overhead, mere white specks against the blue. They may disappear unless the observer has binoculars trained on them. Sometimes they even outclimb the range of the binoculars and disappear altogether. The turkey vulture and the gulls are using thermal soaring.

Thermals are not just odd currents of air rising at random. They have a definite structure that gliders are able to use. When the sun heats the earth's surface, different places give off different amounts of heat. Ploughed fields and concrete roads, for example, are hotter than surrounding woods. Heated air rises quickly from a hot area. Air from the cooler places around it is drawn in, heated, and rises in turn.

Fed by the surrounding air, a swelling bubble of heated air forms. When it reaches a large size, the bubble breaks away from the ground and ascends, all the while being fed by a column of upward rushing air. Moving higher, the bubble breaks loose from the column. It forms a ring of whirling air like a motor car inner tube, with the cooler air flowing up through the middle, then spilling down over the upper ring. The whole ring keeps rising. These invisible inner tubes of circulating air are called thermal shells. Gliding birds use them to ascend thousands of feet above the earth.

To use a thermal, a vulture or other glider first circles to stay within the ascending air inside the ring. The bird adjusts its glide to move downwards at the same speed that the inner air is rising. Thus it can hold its position within the thermal shell and move up with it.

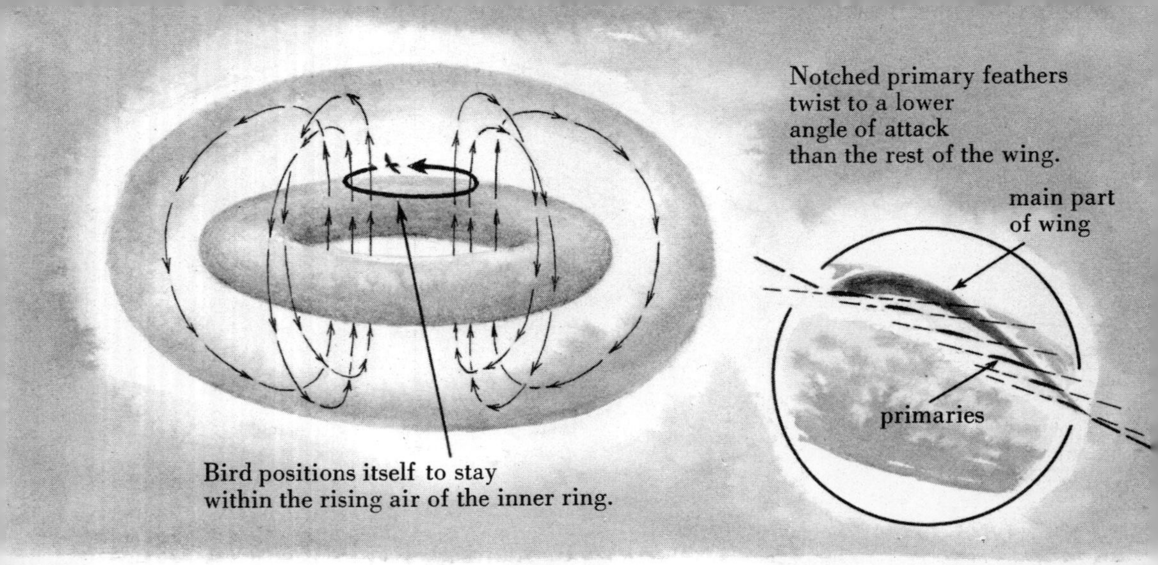

how a bird soars in a thermal shell

The slow-speed, high-lift wings of the turkey vulture are suited ideally to executing the slow, tight circles needed to stay within the thermal shell. The notched primaries, like those of other land soarers, have narrow front webs and wide rear webs. When the vulture glides, the pressure of the airstream twists its primaries so they travel at a lower angle of attack than the rest of the wing. Thus, the thin aerofoil of each primary feather works at its most effective angle. This feature, combined with notching that reduces the wing-tip vortex, and separation between feathers that smooths the entire wing-tip airflow, allows the turkey vulture to glide very slowly at angles of attack up to twenty-eight degrees without stalling. Other vultures, hawks, eagles, and buzzards also have this special wing design for soaring in thermals.

Thermal soaring requires the right conditions, since a bird can use a thermal for soaring only when the air within the shell is rising faster than its own sinking speed. When the sun does not provide enough heat, soaring is not possible. Birds sometimes soar on partly cloudy days, but not under an overcast sky unless the sun has been out earlier. In winter, when the sun's heat is less, thermals are weaker and soaring is more limited.

Even the time of day may determine when a bird will soar. A scientist in India found that the cheel, the vulture with the lightest wing loading and the slowest sinking speed, soars earliest in the morning. The smaller, heavier Indian black vulture, with a faster sinking speed, has to wait until later in the morning. Then the sun has heated the earth more, producing thermals with enough lift so that the heavier birds can soar. The cheel starts soaring much earlier in the day during summer than during winter. Large, heavy birds in India like the crane can soar only in the summer during a few of the hottest hours of the day.

Soaring birds use thermals in a special way to cover long distances when searching for food or during migra-

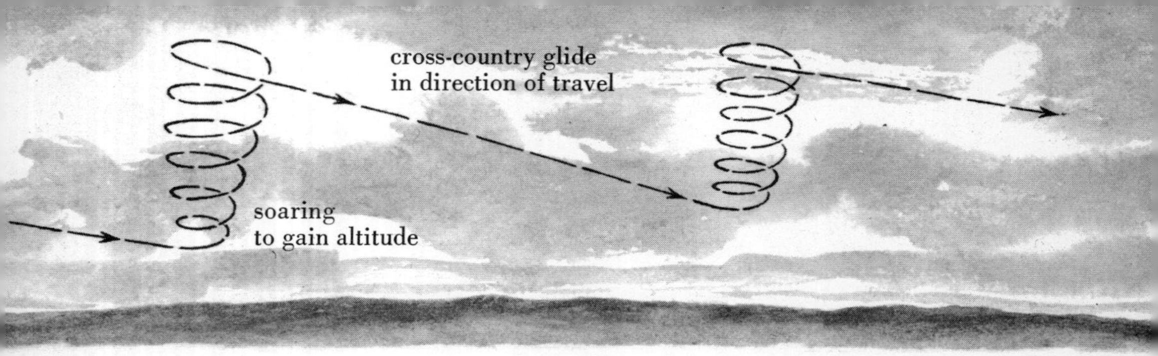

cross-country travel by thermal soaring

tion. First they ride up as high as the thermal will take them. Then, when the ring structure begins to dissolve in the cool, upper air, they glide straight out in their direction of travel. From a starting height of several thousand feet, a bird can glide a long way. As it loses altitude, it finds another thermal, rides up again to regain height, and heads cross-country once more.

Birds also soar over the ocean. Sometimes there are even thermals at sea. During the winter when the ocean temperature is often higher than that of the air above, warm air currents rise from the surface. At such times, in a calm or light wind, seagulls can be seen far from land, circling upward. Scientists are not sure about the form of these currents, but many think that they are upright thermal cells standing side by side like a honeycomb, with warm air rising through their centre. Gulls could soar within these cells or glide across their top.

sea thermals

deflection currents

soaring over the ocean

Gannets, some shearwaters, and certain other seabirds soar on deflection currents pushed up by wind along the slope of wave swells. By continually manoeuvring to take advantage of these updraughts, gannets manage to do quite a bit of soaring. However, wave deflection currents constantly change, are not very strong, and require lots of flapping and agile manoeuvring to shift from one updraught to the next. They are not a dependable enough source of power for the largest sea gliders, the albatrosses.

Another, more reliable power source for ocean soaring is the strong, level flow of the wind itself. Steady, prevailing winds blow across great stretches of the sea, and albatrosses follow these wind patterns in their travels. Winds at sea meet no interruptions, like the hills and valleys on the land, to slow them down, so they blow about twice as fast as land winds. Sea soarers use their long, pointed, high-speed wings to get the most power from these winds.

How can a bird soar on a current blowing straight and level? Take a stack of several playing cards and slide them quickly across a table. You will see that the bottom card moves the least, the next card travels farther, and so on up to the top card, which goes farthest of all. The bottom card stops first, since it receives the most resistance, or friction, from the table's surface. Similarly the lowest layer of wind blowing across the ocean moves the slowest, because the friction is greatest at the surface. The higher the altitude, up to a height between fifty and one hundred feet, the faster the wind blows. Thus, if the wind speed is twenty miles per hour at the surface, it may be forty miles per hour or more at fifty feet. Albatrosses and some shearwaters use these differences in wind speed to soar.

A good example is the wandering albatross of the Southern Ocean, which is the largest and greatest sea soarer of all. Its wings sometimes span over ten feet, and its speed above the waves often exceeds seventy-five miles per hour

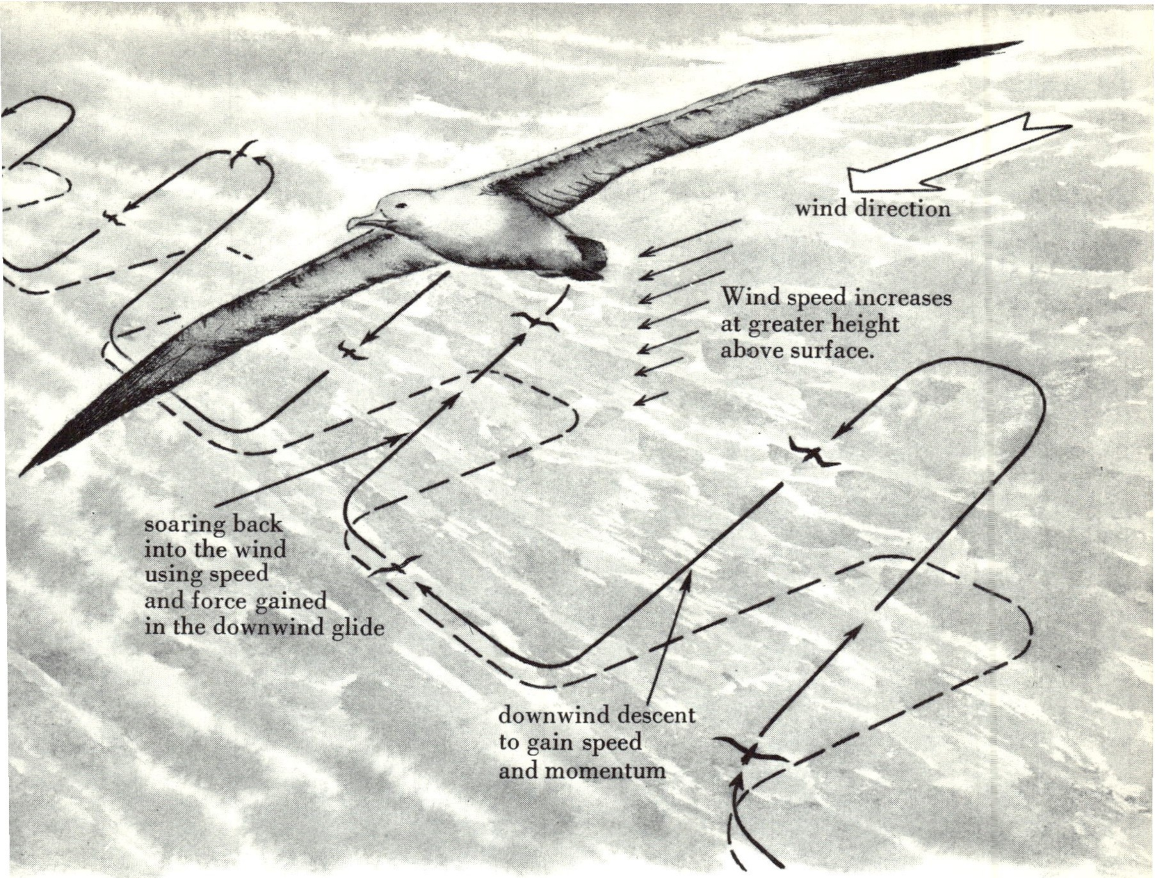

soaring technique of the wandering albatross

in gale force winds above forty miles per hour. When soaring, the wanderer starts out on a steep downwind glide, picking up speed and momentum. As it drops lower and lower it meets air layers moving slower and slower, so in comparison its airspeed builds up faster and faster, and it keeps gaining force. At the bottom of the downwind glide, the albatross turns with a wing tip skimming the swells and climbs back into the wind, using its power gathered downwind to regain height. As it soars upward, the wind layers are moving faster and faster, aiding its lift. Thus, although the albatross is travelling slower over the sea as it mounts upwards, it loses very little airspeed. Once having regained its height, it turns into another downwind glide, and so on, repeating these movements again and again on a zigzag course above the waves. Without flapping and with only small adjustments of its wings, the wandering albatross can travel for hours and days over the vast distances of the sea.

Gliding, the most basic form of flight, is in some ways the most difficult to understand. When birds flap, something physical and tangible can be seen at work. You can sense the muscles pulling and the feathers whirring. But when birds glide there is nothing to see except poised feathered forms gliding on thin air. The seasoned observer feels even more wonder when elegant creatures of the air begin to soar. Motionless and relaxed, they master their environment. They conquer the earth's terrain, harness the sun, tame the wind, and for a while flout the laws of gravity.

6 Flapping and Hovering

The beat of wings fills the air. A crow flies slowly over a field, flapping its black wings up and down, up and down. Calling once or twice, a chaffinch beats quickly, stops momentarily, then beats again as it follows a roller-coaster path back to its nesting tree. A peregrine falcon mounts higher and higher in the sky on swift stroking wings that cut the air like feathered knife blades. A huge grey heron lumbers across a tidal inlet, slowly raising and lowering its broad, majestic surfaces. The ruby-throated hummingbird vibrates its wings in a gentle buzz as it backs away from a columbine and suspends itself in mid air before another flower. All these birds, so different in many ways, depend upon their own muscles for power and upon the constant flapping of their wings to fly.

Many people believe that flapping birds row or swim through the air. Flapping flight is not that simple. A bird does not push backwards in order to move forwards, the way a rower moves his oars or a swimmer uses his hands and feet. A bird's power stroke moves forwards and down, while its backstroke serves mainly to put its wings into position for the next power stroke. Forwards and down, backwards and up. How does this motion propel the bird forwards?

In a gliding wing both the inner wing and the hand are spread wide to form one lifting surface. In a flapping wing, however, each section performs a separate task. The inner wing lifts while the hand section pulls.

Look at the wings of a small flapping flier, the blue tit, in flight. Because its wings are flexible they twist against the airstream on the downstroke. In its hand sections, the inner trailing edge of each primary feather overlaps the leading edge of the feather behind it, forming a con-

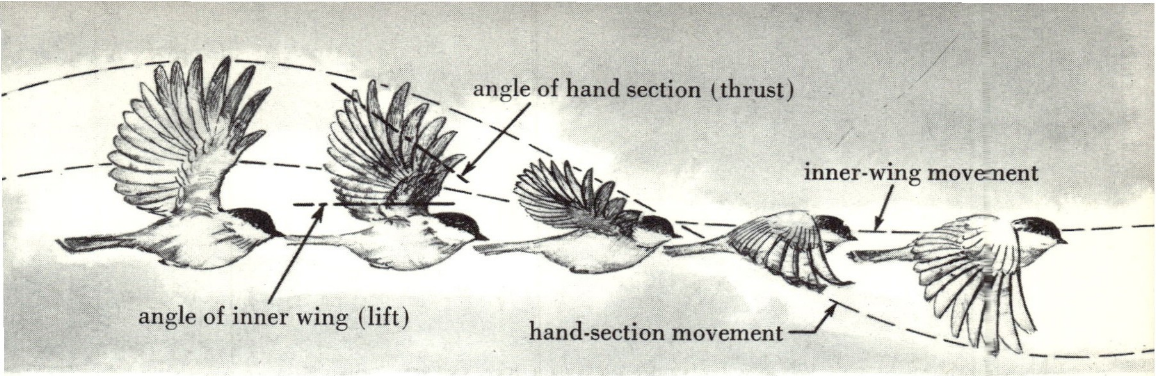

wing movement of tit during downstroke

tinuous aerofoil surface. The hand sections are more flexible than the inner wings. They travel farther during the stroke, so they move faster, meet greater air pressures, twist more, and strike the airstream at a different angle from that of the inner wings. While the tit's inner wings tilt up in front to provide lift, its hand sections tilt down so that their aerofoils create a forward, rather than an upward force. This forward, pulling force is called thrust. As the hands descend, their thrust pulls them forwards, and they in turn pull the inner wings, which lift the bird.

The wing tips of the tit are notched, and the outer end of each primary feather twists separately. The primaries have narrow front webs and wide rear webs, precisely designed outlines, and an aerofoil shape across, so that each feather twists up and backwards at just the proper angle

wing movement of tit during upstroke

owl starting downstroke

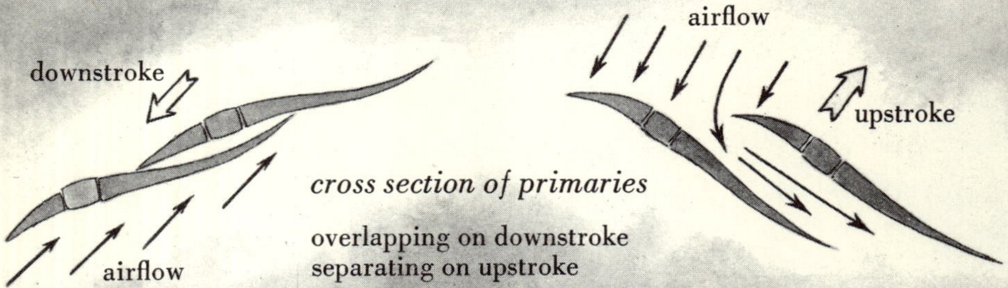

cross section of primaries

overlapping on downstroke
separating on upstroke

upstroke of pelican

Partial folding, or flexing, of wings lessens drag.

on the power stroke to give the most thrust and the least drag. Each primary twists along its own length, so that it forms a continuously changing angle with the airstream as the wing moves up and down. Pointed wing tips, like those of the swallows, work a little differently. There the feather overlap is so complete that the entire wing tip twists as a single unit. In either case, pointed or notched, the whole wing constantly changes shape, curvature, speed, and angle of attack as it moves.

At the end of the downstroke, the wings reverse direction and start moving back and up. On the upstroke, the tit and other small, short-winged birds do little or no work. The inner ends of their primary feathers, which press one on top of the other on the downstroke, separate like slats of a Venetian blind. This change of position allows air to slip through easily and cuts down drag. To make the upstroke even easier, small short-winged birds fold their hand sections. At the top of the stroke they flip their wings out wide again and start the next downstroke. Small, long-winged birds like swallows, as well as larger birds like gulls, keep their hand sections extended on the upstroke. To lessen drag, they flex, or partly fold, their wings. Their hand sections possibly may do some work by pushing backwards as they spread wide for the following stroke.

During the whole stroke cycle, the inner wings maintain their lifting angle. Since they travel a shorter distance up and down, they move more slowly than the hand sections. In fact, as lifting surfaces, they could work just as well without moving at all. The only reason the inner wings do change position is to act as the levers that drive the hand sections. Some larger birds like the American black skimmer, which flies just above the water and dips its lower bill to feed, have such a shallow wing stroke that their inner wings barely move. Many high-speed fliers like the dunlin, have a much deeper stroke, which carries their inner wings through wider arcs.

Swallows, swifts, nightjars, terns, and skimmers wing along so lightly and swiftly that they appear to flap with much less effort than many other flapping birds. All these species feed on the wing and spend much of their time in the air. The secret of their style of flying, beside their efficient long, pointed wings, is that for their body size and weight they have plenty of wing area and, thus, a low wing loading. If a flapping flier has lots of wing area, it can beat its wings more slowly and easily with shorter strokes and still fly at high speed for long periods of time using little energy. Since this type of bird needs so little thrust to provide lift, it can use all the rest to drive itself forward.

The American robin is a migratory species closely related to our fieldfare. It is a fairly hard-working flapper that does not spend much time in flight except during migration, has about thirteen square inches of wing area per ounce of weight. The American nighthawk (related to our nightjar), which flies long hours each day to feed aloft, has about the same body size as the robin, but the much lighter wing loading of twenty square inches per ounce. This difference helps to account for the nighthawk's lazy mastery of flight.

The Leach's petrel, which lives and feeds on the surface of the ocean, has the lightest wing loading—forty-two square inches per ounce—of any flapping flier. This bird hardly seems to work at all as it floats along, bouncing like a butterfly above the waves. At the other end of the scale, the scaup duck has very small wings for its weight, only two and seven-tenths square inches per ounce. The scaup has to use plenty of power, and so flaps fast and hard to stay aloft. Here are some wing loadings for different species of flapping fliers:

species	*weight* (ounces)	*wing area* (square inches)	*wing loading* (square inches per ounce)
ruby-throated hummingbird	.1	1.92	19
willow tit	.44	11.8	26.8
barn swallow	.6	18.4	30.7
American chimney swift	.61	16.1	26.4
Leach's petrel	.94	39	42
American nighthawk	2.7	54	20
American robin	2.9	37.8	13
scaup	27	73.2	2.7
herring gull	30	310	10.7
peregrine falcon	43.2	208	4.8
great blue heron	67.4	687	10.2
Canada goose	200	437	2.2

To control their air speed, flapping fliers can vary the speed, strength, and length of their wing stroke. In general, small birds like tits, warblers and sparrows flap the fastest, medium-sized birds from the size of thrushes to crows flap at moderate speeds, and large birds like the heron flap very slowly. Small birds have to flap fast, because their small wings are not efficient. They must make up in speed of stroke what they lose in edge and tip turbulence. Large birds have efficient wings, but there is another reason for their slow wing beats. Their muscles cannot move the big, broad surfaces of their wings through the air rapidly, so they must flap slowly and use much effort in doing so. Perhaps as a result, many large species that spend much of their time in flight, like condors and albatrosses, have come to rely almost entirely upon gliding.

How fast do birds fly? Precise flight speeds are hard to obtain due to variations in wind speed and direction, as well as the difficulty in knowing whether a bird is cruising or in a hurry when the check is made. Scientists are still trying different methods of measuring, so future research may bring more accurate results. In the meantime, here are some general speed ranges and approximate flight speeds drawn from many field observations:

type	*speed range* (miles per hour)	*individual observations* (varying conditions)
warblers	20–30	American yellow warbler—22
herons and egrets	20–30	snowy egret—30
thrushes	25–35	American robin—33
swallows and martins	30–40	swallow—46
ducks and geese	40–60	tufted duck—66
shorebirds	40–60	American willet—47
falcons	40–60 +	peregrine (diving)—180–200
swifts	60–70 +	Indian swift—170–200

At certain times, such as when pursuing prey or fleeing from an attacker, birds can put on tremendous bursts of speed. Even the crow, normally a slow flier, has been known to fly at sixty miles per hour. A flock of dunlin and other sandpipers once passed an aeroplane that was travelling at ninety miles per hour, and the pilot estimated their air-speed at one hundred and ten miles per hour. Large swifts are believed to be the fastest of all birds. Their normal speed of sixty to seventy miles per hour is surprisingly high, but they have been observed moving at much higher rates. A scientist in India timed two species of swifts in level flight over a two-mile course at speeds between one hundred and seventy and two hundred miles per hour. Even allowing for considerable error, no other bird approaches this speed in level flight, although the peregrine comes out of its dives at high speeds.

How can swifts fly so fast? The structure of one American species, the chimney swift, provides several reasons. It has a deep breastbone, which carries large, powerful flight muscles. The body is bullet-shaped and joined very smoothly to the sharp, slender wings so that the drag is kept very low. The wings themselves are large, giving the bird a light wing loading, and they are almost all hand section, with a very short inner wing. This design means that the swift is practically all propeller, all forward thrust. The long primaries form stiff propeller blades, with thin, flat, high-speed aerofoils beating at a fast ten strokes per second. However, what may be a swift's most unusual feature is its probable two-way power stroke. On the upstroke the rigid, heavy wrists do not appear to bend at all,

American chimney swift

and the primaries do not seem to separate to let the air through. Since its wings do not change form to allow for a recovery stroke, probably unlike other birds the swift develops a good part of its forward thrust on the upstroke. Scientists have not proven that it uses a two-way power stroke, but if this theory is true, the uninterrupted thrust would help explain the wide difference in speed between the swifts and all other birds.

The swift, fastest of all flapping fliers, is related to the slowest, the hummingbird, which can fly at zero miles per hour for extended periods, even in calm air. When the hummingbird suspends itself in mid air to sip a flower, it uses a special form of flapping flight called hovering. It has only six secondary feathers, so in proportion its propeller-like hand sections are even longer than the swift's. Most of its thrust and lift come from the extreme speed of its wings. A ruby-throated hummingbird beats about fifty times a second. To achieve this rate, its flight muscles are enormous for its size, and it burns energy faster than any other bird. The wings hinge on freely swivelling shoulder joints and can tilt at almost any angle.

In level flight a rubythroat probably flies much as a swift does, thrusting forward on both the upstroke and the downstroke. Its wings twist, the outer parts pull, and the

hovering flight of the ruby-throated hummingbird

small inner parts lift. However, when the hummingbird hovers, it tips its body at a high angle and beats its wings back and forth in a horizontal direction. On the forward stroke, the wings thrust straight up. All the force goes straight upward, so there is no forward thrust. On the backstroke, the wings swivel backward at the shoulder to tilt the other way. For an instant the upper surfaces become the lower surfaces, so they keep thrusting straight up and lift the bird like the rotors of a helicopter. By adjusting its thrust to equal its weight, the rubythroat stands still in the air. To move forward, it angles its wing strokes forwards, to back up, it tilts them backwards. A helicopter tilts its rotors in a similar way. Swifts, some warblers, and flycatchers, among other birds, also are able to hover briefly by beating their wings back and forth.

The kingfisher is one of a number of birds that can stand still in flight while searching out their prey. However, it does not really hover. Hummingbirds can hang still as long as necessary in perfectly calm air, while the kingfisher, in contrast, must have at least a light wind. It flies against the wind and slows down by increasing its angle of attack, until it is moving just fast enough to overcome the force of the wind. Although its ground speed, the rate at which the kingfisher is crossing the water, is zero, its airspeed, the rate at which it is moving through the airflow, is equal to the speed of the wind. Without the wind, its wings could not create enough lift to hold it up for more than a brief moment.

From the smallest and fastest to the largest and slowest, beating wings take to the sky. Down and forwards, up and back, twisting, bending, folding, spreading, up and down, up and down they go. With ever-changing curving forms, so complex and subtle they defy the physicist and engineer, flapping wings thrust forward, generate lift, and sustain birds in steady progress through the air.

7 Birds in Flight

A herring gull stands on a beach facing the wind. It unfolds its wings, spreads its tail, and starts to run across the sand, flapping hard. Moving slowly at first, the gull soon picks up speed. Its wings lift, carrying its running feet upward from the ground, and the gull folds them snugly against its body. Now it moves faster and faster, wing tips stroking just above the sand. Soon it reaches full airspeed and starts a shallow climb out over the water. Before long, at about a hundred feet, its wings feel the upsurge of a thermal rising from the beach. It turns sharply to the left and positions itself in the rising air.

After about an hour of soaring, the gull is directly above a favourite feeding spot. To lose altitude quickly, it slides off to one side into a steep, twisting dive. A few feet from the surface, it pulls out of the dive with wings outstretched, tail fanned open, and feet thrust down. After a short, slow glide, it flaps easily a few times and alights gently with a slight splash.

The gull was using the complex patterns of actual flight, its wings and tail spreading, folding, tilting, twisting, always changing, everything happening at once. Flapping, hovering, gliding, and soaring are only the basic methods that birds use to stay up in the air. Each type of flight is made up of many smaller manoeuvres. Even the terms glider and flapper, used to describe different kinds of birds, are not clear-cut and simple. The gull is essentially a glider, but like all other birds it must flap hard to take off and climb. It also flaps at other times when soaring is not possible. On the other hand, the swift, a great flapper, often spreads its wings in long, smooth glides. Sometimes when the wind is light and thermals lift its insect food up high, it performs a limited form of soaring

on fixed wings, flapping only intermittently. Even its basic stroke pattern, like that of many flapping fliers, consists of several beats, then a short glide. In actual flight, birds blend many diverse movements together into a highly complex mixture.

One of the most important individual manoeuvres is the takeoff, such as the one made by the herring gull. On takeoff, the gull faced into the wind, which was blowing at ten miles per hour. Thus, as soon as it opened its wings they were operating in an airflow of ten miles per hour, although it was not even moving over the ground. When the gull ran into the wind, it moved its wings through the air fifteen miles per hour faster, thus increasing the speed of the airflow across them from ten to twenty-five miles per hour. At that point the gull's airspeed was twenty-five miles per hour, although its ground speed was only fifteen. Taking off into the wind allowed it to gain extra lift from the additional ten miles per hour of airflow that the wind provided.

The gull flapped its wings at a high angle of attack to get as much lift as possible while its speed was still low. Its alula feathers smoothed the wing's airflow, and its tail spread wide for extra lifting area. By flapping faster, harder, and with a deeper stroke than it would use in level flight, it got the greatest possible thrust from its hand sections. As it moved faster and faster its inner wings lifted more and more until their force became greater than its weight and drag. The gull was in the air.

Even with help of a head wind, taking off is the most strenuous part of flying, especially for larger birds that have lots of weight to lift. Taking off in calm air is much harder, since a bird must reach full flying speed completely by the power of its own flapping. Pigeons, for example, flap so hard on takeoff in a calm that their wings slap together at the top of the stroke. One reason is that pigeons have a high wing loading, which means less built-in lift and harder flapping on takeoff. The gull and other big-winged birds, like egrets and herons, have lighter wing loadings, so they rise with less effort. When egrets, herons, and some other similar birds take off, they form special slots by overlapping their inner wing trailing edges with the front of their spread tail. This wing position allows their tail to lift at a very high angle.

Different birds increase their speed on takeoff in different ways. Many birds run along the ground. Egrets, herons, and other waders have long legs for jumping off. Mallards and some other ducks push hard with their webbed feet to leap up from the water. Divers, coots, and swans have heavy wing loadings, so they must patter quickly along the surface and flap very hard to gain sufficient takeoff speed.

Birds that perch in trees have little trouble taking off. A warbler is in the air the moment it leaves a branch. If it dips slightly before reaching its flying speed, it has plenty of room to rise up again.

The same is true of cliff-dwelling birds like the gannet. A gannet simply walks off its front porch into the air. However, gannets may have other problems. They catch fish by diving underwater from high in the air. Sometimes when they surface, the wind has calmed down. They run and flap hard, trying to take off, but their glider wings with small hand sections and light flight muscles limit their power. They have to wait for a breeze. One observer reported the destruction of hundreds of gannets that were carried onshore in a heavy surf when they were unable to rise from the water.

gannets

The largest gliders of all, the condors and albatrosses, need a long takeoff run any time the breeze is not strong. When threatened on the ground, they will even throw up recently eaten food to lighten themselves for quicker take-off. Condors nest on high mountain cliffs where they glide downward on takeoff. In South America the people trap them by baiting narrow wooden corrals. Once they land inside, the giant birds do not have enough room for a take-off run.

After its takeoff, the gull performed another basic manoeuvre—climbing. Once it had reached flying speed, it climbed by holding its wings at a higher angle of attack than it would use in level flight. The gull also flapped harder for greater thrust and lift. As on takeoff, it rose higher because its lift and thrust exceeded its weight and drag.

The gull controlled its angle of climb by the amount of lift it produced and also by adjusting the angle of its tail surfaces. Its body can tilt up or down, around the balancing point of its weight, called the centre of gravity. Raising the rear of its tail makes the airflow strike the upper surface,

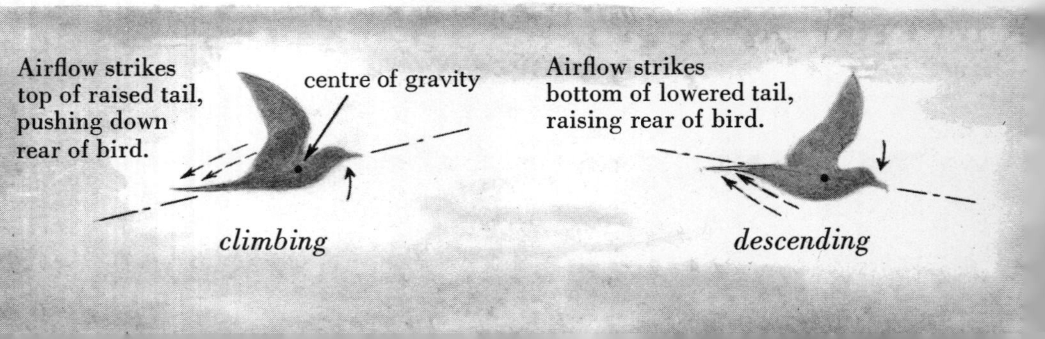

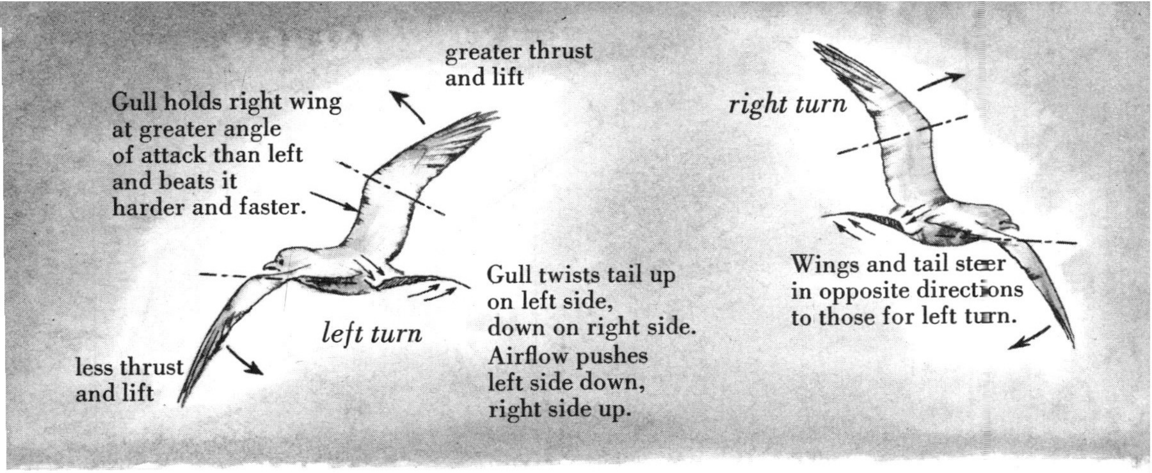

forces its tail down, and tilts its wing at a higher angle for climbing. In a similar way, lowering the tail lowers the wing angle and the gull levels off or heads downward. Thus its tail acts as a horizontal control surface that holds its wing set at the proper angle. In addition, the gull can make momentary adjustments of its wing angle during takeoff or landing by rotating its wing at the shoulder joint.

Turning is another important manoeuvre. To turn left, the gull flapped its right wing faster and harder than its left and twisted its right leading edge upwards to increase the angle of attack on its right side. The increased thrust and lift there pulled its right wing forwards faster than its left and also raised it higher. The gull's left wing, beating more easily at a lower angle, had less thrust and lift, so it fell behind and below the right wing. The right wing pivoted forwards, tipping or banking the gull towards the inside of the turn. Banking enables a bird to make a smooth turn with complete control and prevents skidding. It is somewhat similar to the way you lean when you make a turn on a bicycle.

The gull also used its tail to turn, first spreading it wide for greater controlling effect. Then it raised the rear of its tail on its left side, so the airflow striking its upper surface pushed it down. It lowered the right rear side of its tail so the air striking from below raised it. Twisting its tail this way caused the gull to swing more sharply to the left. To turn to the right, it twists its wings and tail in the opposite way and beats its left wing harder than its right. To turn while gliding, it twists its outstretched wings and its tail in the same direction that it does when flapping. A gliding turn is usually slower and flatter than a flapping turn, because in gliding there is no thrust to pull the outside wing around quickly.

Birds with large tails are generally more adept at close manoeuvring than birds with small tails. Long, forked tails, like those of the swallow and some terns, are especially effective in high-speed turns, since their large surface area far from the body gives them plenty of leverage. On the other hand, fast birds with small tails, like the swift, rely largely upon wing stroke and twist for turning.

When the gull was over its landing spot, it performed another frequent manoeuvre—diving. It dived by first partly folding, or flexing, its wings to decrease their lift and to move back their centre of lift, which is normally well in front of the centre of gravity. In a bird's normal position the centre of lift tends to push it up in front. Shifted farther back, it tends to nose it forwards into a dive. The gull also tilted the rear of its tail down to push its wings at a downward angle. To steer in a downward spiral, it twisted its wings and tail. Since the gull was gliding, gravity alone pulled it down. Birds as large as a gull usually glide when they dive, while many smaller birds flap to drive themselves along. To speed up the dive, the gull can fold its wings more and descend at a steeper angle.

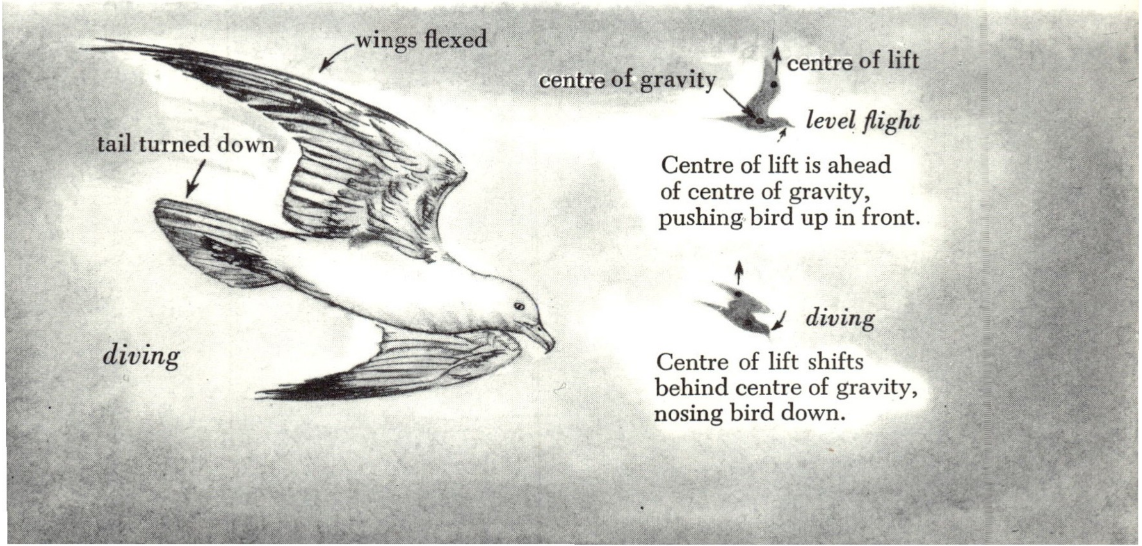

Even slow gliders can travel very fast in a dive. For example, a large lammergeyer, or bearded vulture, was once clocked from an aeroplane diving alongside at one hundred and ten miles per hour. The fastest and greatest diver is the peregrine falcon. With its sharp wings flexed beside its sleek powerful body, it often makes long dives while hunting. The peregrine has been known to overtake and easily pass an aeroplane diving at one hundred and seventy-five miles per hour as it aimed for some ducks flying far below.

The gull slowed down near the end of its dive by extending its wings slightly for greater lift, and by tilting its tail up to increase its angle of attack. Finally, to pull out of the dive, the gull spread its wings wide for maximum lift, opened its tail like a braking parachute, and lowered its legs to increase drag and cut speed further as it glided to a landing.

Landing, like takeoff, is an essential manoeuvre. The gull, like all birds, lands as slowly as possible to keep from injuring itself and to reach a certain spot accurately. During its landing the gull headed into the wind to decrease its ground speed while still maintaining enough airspeed for lift. It flew as slowly as possible at a high angle of attack, with alula and tail feathers spread for maximum

lift and braking effect. By fanning its wings back and forth horizontally, it directed their thrust and lift upwards and backwards to check its forward descent. Finally, moving very slowly just on the verge of stalling, it touched down.

A light wing loading allows the gull and many other birds to make slow, gentle landings. Herons have to protect their long slender legs when they land at their tree nests, so the ability to land slowly is especially important to them. Small tree birds that alight on thin branches, often near nests full of fragile eggs, must land delicately and accurately. Their legs bend at the joints like shock absorbers and take up the last bit of speed. Woodpeckers and other tree-trunk feeders lose speed by making a final upswing, so that they land in a vertical position.

On open ground, birds like the heron often land faster and run to a stop. Running on landing is also the method used by many shorebirds and other fast fliers that live in the open. Water is certainly the easiest place to land. A little too much speed or a slight error in the approach merely means a bigger splash. As a result, ducks, geese, and other waterbirds with high wing loadings can land very fast and skid to a stop with their feet thrust forward as brakes.

The basic flight manoeuvres used by the gull are much the same as those used by all birds. The landing technique of an eagle is not very different from that of a robin. The same fundamental forces of lift, thrust, weight, and drag are at work when any bird takes to the air.

A knowledge of flight characteristics can improve identification of birds in the field, especially when long distance or poor light obscures their markings and colours. Field guides contain many flight notes that can help you distinguish different species. The speed of wing beats tells you at once the general size of a bird, since small birds flap quickly and large birds flap slowly. Instead of flapping steadily, certain species flap, pause, and flap. Some of them maintain a level flight path by gliding between flaps. Others fold their wings when not flapping, dipping down and rising again with the change in their lift. Some birds, like the common sandpiper, use a short, fluttering wing stroke that sets them apart. Even the more subtle qualities

of the wing stroke can be distinctive. The smooth, effortless stroke of the golden plover is quite different from the hard driving beat of the peregrine falcon.

Study the different flight formations. Canada geese, golden plovers, and many waders and ducks fly in V-shaped groups. Eider ducks appear in long single lines just above the waves. Dunlin twist and turn together in tight flocks.

Learn the flight silhouettes of hawks, buzzards, and eagles, which usually are seen soaring far overhead. Flight profiles also can help you identify them when they are turning in a distant thermal.

In addition, with a knowledge of bird flight you can better understand and appreciate how birds live and fly in the ever-changing conditions of season, wind, and weather. The most fascinating thing about birds is not their pattern or colour, but their absolute mastery of the air. In the vast realm of nature, few things can compare with the spectacle of birds in flight.

Bibliography

Aymar, Gordon C. *Bird Flight*. London: Allen Lane, 1936.

An older work which has much valuable material and especially fine photographs.

Cone, Clarence D. "The Soaring Flight of Birds." *Scientific American*, April 1962, pp. 130-140.

A comprehensive study of the different kinds of soaring.

Greenewalt, Crawford H. *Hummingbirds*. Garden City, New York: Doubleday and Company, Inc., 1960.

Contains an excellent explanation and photographs of hummingbird flight.

Jameson, William. *The Wandering Albatross.* London: Rupert Hart-Davis Ltd., 1959.

Chapter four thoroughly examines wind layer soaring.

Peterson, Roger Tory. *A Field Guide to the Birds of Britain and Europe.* London: William Collins Sons & Co. Ltd.

A reliable field guide with essential notes and silhouettes for recognizing birds in flight.

Robbins, Chandler S.; Brunn, Bertel; and Zim, Herbert S. *Birds of North America.* London: Paul Hamlyn Ltd, 1966. (P)

Another valuable field guide with many flight notes and pictures.

Storer, John H. *The Flight of Birds.* Bloomfield Hills, Michigan: Cranbrook Institute of Science, Bulletin No. 28, 1948.

An invaluable book for the serious student of bird flight. Contains many sequence photographs from slow-motion films.

Van Tyne, Josselyn, and Berger, Andrew J. *Fundamentals of Ornithology.* London: Chapman & Hall Ltd., 1959.

Has a fine chapter and bibliography on bird flight.

Welty, Joel Carl. *The Life of Birds.* London: W. B. Saunders Co. Ltd., 1963.

See chapters one, four, twenty-nine, and the extensive bibliography.

Index

** Indicates illustration*